MEMOIRS
of the
American Mathematical Society

ISSN 0065-9266 (print)
ISSN 1947-6221 (online)

Number 1511

AF846516

Symbolic Dynamics for Nonuniformly Hyperbolic Maps with Singularities in High Dimension

Ermerson Araujo
Yuri Lima
Mauricio Poletti

September 2024 • Volume 301 • Number 1511 (fifth of 7 numbers)

AMS AMERICAN MATHEMATICAL SOCIETY

Library of Congress Cataloging-in-Publication Data

Cataloging-in-Publication Data has been applied for by the AMS.
See http://www.loc.gov/publish/cip/.
DOI: https://doi.org/10.1090/memo/1511

Memoirs of the American Mathematical Society

This journal is devoted entirely to research in pure and applied mathematics.

Subscription information. Beginning in 2024, Memoirs will be published monthly through 2026. Memoirs is also accessible from www.ams.org/journals. The 2024 subscription begins with volume 293 and consists of twelve mailings, each containing one or more numbers. Individual subscription prices for 2024 are as follows. For electronic only: US$1013. For paper delivery: US$1,163. Add US$22 for delivery within the United States; US$150 for surface delivery outside the United States. Upon request, subscribers to paper delivery of this journal are also entitled to receive electronic delivery. For information on institutional pricing, please visit `https://www.ams.org/publications/journals/subscriberinfo`. Subscription renewals are subject to late fees. See `www.ams.org/journal-faq` for more journal subscription information. Each number may be ordered separately; please specify number when ordering an individual number.

Back number information. For back issues see `www.ams.org/backvols`.

Subscriptions and orders should be addressed to the American Mathematical Society, P. O. Box 845904, Boston, MA 02284-5904 USA. *All orders must be accompanied by payment.* Other correspondence should be addressed to 201 Charles Street, Providence, RI 02904-2213 USA.

Memoirs of the American Mathematical Society (ISSN 0065-9266 (print); 1947-6221 (online)) is published bimonthly (each volume consisting usually of more than one number) by the American Mathematical Society at 201 Charles Street, Providence, RI 02904-2213 USA. Periodicals postage paid at Providence, RI. Postmaster: Send address changes to Memoirs, American Mathematical Society, 201 Charles Street, Providence, RI 02904-2213 USA.

This publication is indexed in *Mathematical Reviews®*, *Zentralblatt MATH*, *Science Citation Index®*, *Science Citation Index™-Expanded*, *ISI Alerting Services℠*, *SciSearch®*, *Research Alert®*, *CompuMath Citation Index®*, *Current Contents®/Physical, Chemical & Earth Sciences*.
This publication is archived in *Portico* and *CLOCKSS*.
Printed in the United States of America.

∞ The paper used in this book is acid-free and falls within the guidelines established to ensure permanence and durability.
Visit the AMS home page at `https://www.ams.org/`

10 9 8 7 6 5 4 3 2 1 29 28 27 26 25 24

Contents

Abstract

We construct Markov partitions for non-invertible and/or singular nonuniformly hyperbolic systems defined on higher dimensional Riemannian manifolds. The generality of the setup covers classical examples not treated so far, such as geodesic flows in closed manifolds, multidimensional billiard maps, and Viana maps, and includes all the recent results of the literature. We also provide a wealth of applications.

Received by the editor November 22, 2021, and, in revised form, April 2, 2022.

Article electronically published on September 23, 2024.

DOI: https://doi.org/10.1090/memo/1511

2020 *Mathematics Subject Classification.* Primary 37-02, 37B10, 37C05, 37C35, 37C83, 37D25, 37D35.

Key words and phrases. Markov partition, Pesin theory, symbolic dynamics.

The authors thank Viviane Baladi, Snir Ben Ovadia, Jérôme Buzzi, Gianluigi Del Magno, Mark Demers, François Ledrappier, Roberto Markarian, Omri Sarig, Domokos Szász for useful explanations and suggestion. Araujo is supported by CAPES. Lima is supported by CNPq and Instituto Serrapilheira, grant "Jangada Dinâmica: Impulsionando Sistemas Dinâmicos na Região Nordeste". During the preparation of this manuscript, Poletti was supported by the ERC project 692925 NUHGD.

CHAPTER 1

Introduction

In 2013, Sarig constructed countable Markov partitions with full topological entropy for $C^{1+\beta}$ surface diffeomorphisms [**Sar13**]. Since then, his method has been further developed to other classes of systems, including classes where the classical literature of Markov partitions was not able to handle, such as billiard maps. Here are some of the recent developments:

- Lima and Sarig for three dimensional flows without fixed points [**LS19**].
- Lima and Matheus for surface maps with discontinuities [**LM18**].
- Ben Ovadia for diffeomorphisms in any dimension [**BO18**].
- Lima for one-dimensional maps [**Lim20**].

The goal of the present work is to go far beyond the above results, constructing Markov structures for maps on higher dimensional Riemannian manifolds, with the novelty of allowing the map to be noninvertible and have a singular set, and the manifold to be disconnected and noncompact. This unified framework allows us to cover many classical examples of nonuniform hyperbolic dynamics, such as geodesic flows in closed manifolds, billiard maps, and Viana maps. Also, this wider framework includes the four works cited above, in some cases providing even better results (in comparison to [**Lim20**]). Not only this: using the recent work of Ben Ovadia [**BO21**], we are able to identify the points coded by recurrent sequences.

Before stating the main result, let us mention some of its applications. Let N be a closed C^∞ Riemannian manifold, let X be a $C^{1+\beta}$ vector field on N s.t. $X_x \neq 0$ for all $x \in N$, and let $g = \{g^t\}_{t\in\mathbb{R}} : N \to N$ be the flow determined by X. Assume that g has positive topological entropy, call it h. Let μ be a g-invariant Borel probability measure on N. We call μ *hyperbolic* if all of its Lyapunov exponents are nonzero, except for the zero exponent in the flow direction. A *simple closed orbit* of length ℓ is a parametrized curve $\gamma(t) = g^t(x)$, $0 \leq t \leq \ell$, s.t. $\gamma(0) = \gamma(\ell)$ and $\gamma(0) \neq \gamma(t)$ for $0 < t < \ell$. Denoting the *trace* of γ by the set $\{\gamma(t) : 0 \leq t \leq \ell\}$, we let $[\gamma]$ denote the equivalence class of the relation $\gamma_1 \sim \gamma_2 \Leftrightarrow \gamma_1, \gamma_2$ have equal lengths and traces, and let $\mathrm{Per}_T(g) := \#\{[\gamma] : \gamma \text{ is simple and } \ell(\gamma) \leq T\}$.

THEOREM 1.1. *Let g be as above. If g has a hyperbolic measure of maximal entropy, then $\exists C > 0$ s.t. $\mathrm{Per}_T(g) \geq Ce^{hT}/T$ for all T large enough.*

The second application deals with equilibrium measures, see Chapter 8 for its definition and Chapter 10 for the definition of the geometric potential J of a flow and its equilibrium measures.

THEOREM 1.2. *For g as above, let $\varphi : N \to \mathbb{R}$ be a Hölder continuous potential or $\varphi = tJ$ for $t \in \mathbb{R}$. Then every equilibrium measure of φ has at most countably many ergodic hyperbolic components, and each of them is Bernoulli up to a period.*

The above result is the higher dimensional extension of [**LLS16**, Theorems 1.1 and 1.4]. Observe that, by Theorem 1.2, if we know that the equilibrium measure is mixing, then it is necessarily Bernoulli. This observation applies for geodesic flows of closed rank one manifolds, where Burns et al. proved that if the pressure gap $P(\mathrm{Sing}, \varphi) < P(\varphi)$ holds, then φ has a unique equilibrium measure and it is hyperbolic [**BCFT18**], and Call and Thompson proved that this measure has the K property and that the unique measure of maximal entropy is Bernoulli [**CT22**]. Here, Sing is the set of vectors with rank larger than one. We thus obtain the following corollary.

COROLLARY 1.3. *Let g be the geodesic flow of a closed rank one manifold N. Let $\varphi : T^1N \to \mathbb{R}$ be Hölder continuous or of the form $\varphi = tJ$ for $t \in \mathbb{R}$. If $P(\mathrm{Sing}, \varphi) < P(\varphi)$, then the unique equilibrium state of φ is Bernoulli.*

In the above context, Knieper proved that the measure of maximal entropy is unique [**Kni98**], and Babillot proved that it is mixing [**Bab02**]. Using these results, we obtain a proof of Corollary 1.3 for $\varphi = 0$ that is different from and does not use [**CT22**]. The proofs of Theorems 1.1 and 1.2 are in Chapter 10, which also contains a theorem that provides a symbolic coding for the considered flows.

The next application deals with multidimensional billiards. Let f be the billiard map associated to one of the following contexts: dispersing billiards with finitely many disjoint scatterers, billiards of hard balls systems, or a three-dimensional Bunimovich stadium as considered in [**BDM06**]; see Figure 1.1. In all of these cases, it is known that the natural smooth f-invariant measure μ_{SRB} is hyperbolic, mixing, and has Kolmogorov–Sinaĭ entropy $h > 0$. Let $\mathrm{Per}_n(f)$ denote the number of periodic orbits of period n for f.

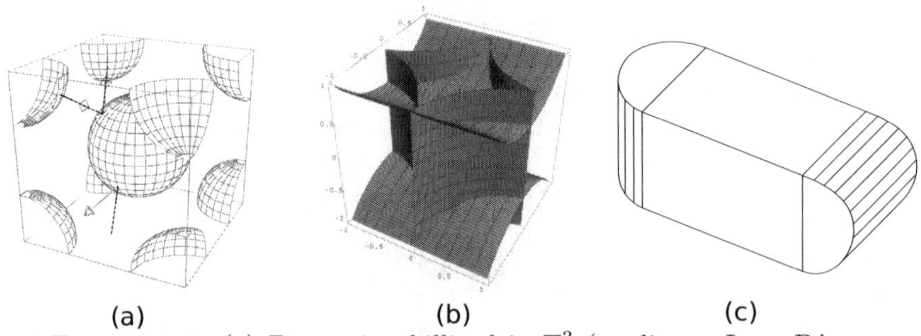

FIGURE 1.1. (a) Dispersing billiard in $\mathbb{T}^3$ (credits to Imre Péter Tóth). (b) Billiard of a hard balls system in $\mathbb{T}^3$ (credits to [**PH00**]). (c) Three-dimensional Bunimovich stadium.

THEOREM 1.4. *For the above billiards, there exists $C > 0$ s.t. $\mathrm{Per}_n(f) \geq Ce^{hn}$ for all n large enough.*

The proof of Theorem 1.4 and a detailed discussion on multidimensional billiards is in Chapter 11. The last applications deal with *Viana maps*, introduced in [**Via97**]. These maps are defined as follows. Let $a_0 \in (1, 2)$ be a parameter s.t. $t = 0$ is preperiodic for the quadratic map $t \mapsto a_0 - t^2$. For fixed $d \geq 16$ and $\alpha > 0$, the associated Viana map is the skew product $f = f_{a_0,d,\alpha} : \mathbb{S}^1 \times \mathbb{R} \to \mathbb{S}^1 \times \mathbb{R}$ defined by $f(\theta,t) = (d\theta, a_0 + \alpha\sin(2\pi\theta) - t^2)$. If $\alpha > 0$ is small enough, then there is a compact interval $I_0 \subset (-2,2)$ such that $f(\mathbb{S}^1 \times I_0) \subset \mathrm{int}(\mathbb{S}^1 \times I_0)$. Hence f has an attractor inside $\mathbb{S}^1 \times I_0$, and so we consider the restriction of f to $\mathbb{S}^1 \times I_0$. Observe that f has a critical set $\mathbb{S}^1 \times \{0\}$, where df is noninvertible. When $\alpha > 0$ is small, f possesses a unique invariant probability measure μ_{SRB} that is absolutely continuous with respect to Lebesgue [**Alv00**, **AV02**], nonuniformly expanding [**Via97**], and mixing [**ALP05**]. As a matter of fact, all these properties hold for a C^3 neighborhood of f. Letting $\mathrm{Per}_n(f)$ as before and $h = h_{\mu_{\mathrm{SRB}}}(f) > 0$, our main theorem implies the following results.

THEOREM 1.5. *If $\alpha > 0$ is small, then $\exists C > 0$ s.t. $\mathrm{Per}_n(f) \geq Ce^{hn}$ for all n large enough. The same applies to a C^3 neighborhood of f.*

THEOREM 1.6. *If $\alpha > 0$ is small, there exists a C^3 neighborhood $\mathscr{U}$ of f such that every map in $\mathscr{U}$ has at most countably many ergodic measures of maximal entropy, and the lift to the natural extension of each of them is Bernoulli up to a period.*

The proofs of Theorems 1.5 and 1.6 are given in Chapter 12.

1.1. Statement of main theorem

To cover all the above contexts at the same time, our framework requires seven assumptions to be satisfied: the first four are assumptions on the manifold, and the last three are on the map. We remark that the first four assumptions coincide with those in [**LM18**], and the last three are similar to those in [**Lim20**]. In the sequel we specify them. Let M be a smooth, possibly disconnected, and/or with boundary and/or not compact, Riemannian manifold with finite diameter. We will always assume that the diameter of M is smaller than one (just multiply the metric by a small constant). We allow M to have infinitely many connected components, as long as it is contained in a larger Riemannian manifold.[1] We fix hence and for all a closed set $\mathscr{D} \subset M$, which will denote the set of *discontinuities* of the map. Given $x \in M$, let T_xM denote the tangent space of M at x. For $r > 0$, let $B_x[r] \subset T_xM$ denote the open ball with center 0 and radius r. Given $x \in M\backslash\mathscr{D}$, let $\mathrm{inj}(x)$ be the *injectivity radius* of M at x, and let $\exp_x : B_x[\mathrm{inj}(x)] \to M$ be the *exponential map* at x.

The Riemannian metric on M induces a Riemannian metric on the tangent bundle TM, called the *Sasaki metric*, see, e.g., [**BMW12**, §2]. Denote the Sasaki metric by $d_{\mathrm{Sas}}(\cdot,\cdot)$, and denote the Sasaki metric on $TB_x[r]$ by the same notation (the context will be clear in which space we are). For nearby small vectors, the Sasaki metric is almost a product metric as follows. Given a geodesic γ joining y to x, let $P_\gamma : T_yM \to T_xM$ be the parallel transport along γ. If $v \in T_xM$, $w \in T_yM$ then $d_{\mathrm{Sas}}(v,w) \asymp d(x,y) + \|v - P_\gamma w\|$ as $d_{\mathrm{Sas}}(v,w) \to 0$, see [**BMW12**, Appendix A]. The properties required on M are about the exponential maps. Here are the first two. For $x \in M$ and $r > 0$, let $B(x,r) \subset M$ denote the open ball with center x and radius r.

REGULARITY OF $\exp_x$: $\exists a > 1$ s.t. for all $x \in M\backslash\mathscr{D}$ there is $d(x,\mathscr{D})^a < \mathfrak{d}(x) < 1$ s.t. for $\mathfrak{B}_x := B(x, 2\mathfrak{d}(x))$ the following holds:

[1] This avoids pathological situations in which natural conditions, e.g. precompactness, are lost.

(A1) If $y \in \mathfrak{B}_x$ then $\text{inj}(y) \geq 2\mathfrak{d}(x)$, $\exp_y^{-1} : \mathfrak{B}_x \to T_yM$ is a diffeomorphism onto its image, and $\frac{1}{2}(d(x,y) + \|v - P_{y,x}w\|) \leq d_{\text{Sas}}(v,w) \leq 2(d(x,y) + \|v - P_{y,x}w\|)$ for all $y \in \mathfrak{B}_x$ and $v \in T_xM, w \in T_yM$ s.t. $\|v\|, \|w\| \leq 2\mathfrak{d}(x)$, where $P_{y,x} := P_\gamma$ for the length minimizing geodesic γ joining y to x.

(A2) If $y_1, y_2 \in \mathfrak{B}_x$ then $d(\exp_{y_1} v_1, \exp_{y_2} v_2) \leq 2d_{\text{Sas}}(v_1, v_2)$ for $\|v_1\|$, $\|v_2\| \leq 2\mathfrak{d}(x)$, and $d_{\text{Sas}}(\exp_{y_1}^{-1} z_1, \exp_{y_2}^{-1} z_2) \leq 2[d(y_1,y_2)+d(z_1,z_2)]$ for $z_1, z_2 \in \mathfrak{B}_x$ whenever the expression makes sense. In particular $\|d(\exp_x)_v\| \leq 2$ for $\|v\| \leq 2\mathfrak{d}(x)$, and $\|d(\exp_x^{-1})_y\| \leq 2$ for $y \in \mathfrak{B}_x$.

The next two assumptions are on the regularity of the derivative $d\exp_x$. For $x, x' \in M \backslash \mathscr{D}$, let $\mathscr{L}_{x,x'} := \{A : T_xM \to T_{x'}M : A \text{ is linear}\}$ and $\mathscr{L}_x := \mathscr{L}_{x,x}$. Observe that the parallel transport $P_{y,x}$ considered in (A1) is in $\mathscr{L}_{y,x}$. Given $y \in \mathfrak{B}_x$, $z \in \mathfrak{B}_{x'}$ and $A \in \mathscr{L}_{y,z}$, let $\widetilde{A} \in \mathscr{L}_{x,x'}$, $\widetilde{A} := P_{z,x'} \circ A \circ P_{x,y}$. By definition, $\widetilde{A}$ depends on x, x' but different basepoints define a map that differs from $\widetilde{A}$ by pre and post composition with isometries. In particular, $\|\widetilde{A}\|$ does not depend on the choice of x, x'. If $A_i \in \mathscr{L}_{y_i,z_i}$ then $\|\widetilde{A_1} - \widetilde{A_2}\|$ does depend on the choice of x, x', but if we change the basepoints x, x' to w, w' then the respective differences differ by precompositions and postcompositions with norm of the order of the areas of the geodesic triangles formed by x, w, y_i and by x', w', z_i, which will be negligible to our estimates. Define the map

$$\tau = \tau_x : \mathfrak{B}_x \times \mathfrak{B}_x \to \mathscr{L}_x \quad \text{by} \quad \tau(y,z) = \widetilde{d(\exp_y^{-1})_z},$$

where we use the identification $T_v(T_yM) \cong T_yM$ for all $v \in T_yM$.

REGULARITY OF $d\exp_x$:

(A3) If $y_1, y_2 \in \mathfrak{B}_x$ then $\|\widetilde{d(\exp_{y_1})_{v_1}} - \widetilde{d(\exp_{y_2})_{v_2}}\| \leq d(x,\mathscr{D})^{-a} d_{\text{Sas}}(v_1, v_2)$ for all $\|v_1\|, \|v_2\| \leq 2\mathfrak{d}(x)$, and $\|\tau(y_1,z_1) - \tau(y_2,z_2)\| \leq d(x,\mathscr{D})^{-a}[d(y_1,y_2) + d(z_1,z_2)]$ for all $z_1, z_2 \in \mathfrak{B}_x$.

(A4) If $y_1, y_2 \in \mathfrak{B}_x$ then the map $\tau(y_1,\cdot) - \tau(y_2,\cdot) : \mathfrak{B}_x \to \mathscr{L}_x$ has Lipschitz constant $\leq d(x,\mathscr{D})^{-a} d(y_1,y_2)$.

Conditions (A1)–(A2) say that the exponential maps and their inverses are well-defined and have uniformly bounded Lipschitz constants in balls of radii $d(x,\mathscr{D})^a$. Condition (A3) controls the Lipschitz constants of the derivatives of these maps, and condition (A4) controls the Lipschitz constants of their second derivatives. Here are some situations when (A1)–(A4) are satisfied:

- The curvature tensor R of M is globally bounded, e.g. when M is the phase space of a billiard map.
- $R, \nabla R, \nabla^2 R, \nabla^3 R$ grow at most polynomially fast with respect to the distance to $\mathscr{D}$, e.g. when M is a moduli space of curves equipped with the Weil-Petersson metric [**BMW12**].

Now we discuss the assumptions on the map. Consider a map $f : M\backslash\mathscr{D} \to M$. We assume that f is differentiable at every point $x \in M\backslash\mathscr{D}$, and we let $\mathscr{C} = \{x \in M\backslash\mathscr{D} : df_x \text{ is not invertible}\}$ be the *critical set* of f. We assume that $\mathscr{C}$ is a closed subset of M.

SINGULAR SET $\mathscr{S}$: The *singular set of* f is $\mathscr{S} := \mathscr{C} \cup \mathscr{D}$.

The singular set $\mathscr{S}$ is closed. We assume that f satisfies the following properties.

REGULARITY OF f: $\exists\mathfrak{K} > 1$ s.t. for all $x \in M$ with $x, f(x) \notin \mathscr{S}$ there is $\min\{d(x,\mathscr{S})^a,\ d(f(x),\mathscr{S})^a\} < \mathfrak{r}(x) < 1$ s.t. for $D_x := B(x, 2\mathfrak{r}(x))$ and $E_x := B(f(x), 2\mathfrak{r}(x))$ the following holds:

(A5) The restriction of f to D_x is a diffeomorphism onto its image; the inverse branch of f taking $f(x)$ to x is a well-defined diffeomorphism from E_x onto its image.

(A6) For all $y \in D_x$ it holds $d(x,\mathscr{S})^a \leq \|df_y\| \leq d(x,\mathscr{S})^{-a}$; for all $z \in E_x$ it holds $d(x,\mathscr{S})^a \leq \|dg_z\| \leq d(x,\mathscr{S})^{-a}$, where g is the inverse branch of f taking $f(x)$ to x.

(A7) For all $y, z \in D_x$ it holds $\|\widetilde{df_y} - \widetilde{df_z}\| \leq \mathfrak{K}d(y,z)^\beta$; for all $y, z \in E_x$ it holds $\|\widetilde{dg_y} - \widetilde{dg_z}\| \leq \mathfrak{K}d(y,z)^\beta$, where g is the inverse branch of f taking $f(x)$ to x.

Although technical, conditions (A5)–(A7) hold in most cases of interest, e.g. if $\|df^{\pm 1}\|, \|d^2 f^{\pm 1}\|$ grow at most polynomially fast with respect to the distance to $\mathscr{S}$.

Remark 1.7. In practice, it is enough to check assumptions (A3), (A4), and (A6) up to multiplicative constants. As a matter of fact, if they hold with multiplicative constants, then we can multiply the metric by a small parameter and increase a to obtain (A3), (A4) and (A6) as stated above. A similar reduction occurs in (A7): if $\|\widetilde{df_y} - \widetilde{df_z}\| \leq \mathfrak{K}d(x,\mathscr{S})^{-c}d(y,z)^\beta$ for some $c > 0$, then we can replace a by $\max\{a, 2c/\beta\}$, β by $\beta/2$, and $\mathfrak{K}$ by $4^{\beta/2}\mathfrak{K}$ and obtain the first inequality in (A7).

Finally, we define the measures that we are able to code.

f-ADAPTED MEASURE: An f-invariant probability measure μ on M is said to be *f-adapted* if $\log d(x,\mathscr{S}) \in L^1(\mu)$. A fortiori, $\mu(\mathscr{S}) = 0$.

Due to assumption (A6), if μ is f-adapted then the conditions of the non-invertible version of the Oseledets theorem are satisfied. Therefore, if μ is f-adapted then for μ-a.e. $x \in M$ the Lyapunov exponent $\chi(v) = \lim_{n\to+\infty} \frac{1}{n}\log\|df^n(v)\|$ exists for all $v \in T_xM$. Among the adapted measures, we consider the hyperbolic ones. Fix $\chi > 0$.

χ-HYPERBOLIC MEASURE: An f-invariant probability measure μ is called *χ-hyperbolic* if for μ-a.e. $x \in M$ we have $|\chi(v)| > \chi$ for all $v \in T_xM$.

The next theorem is the main result of this paper. For that, let $\widehat{f} : \widehat{M} \to \widehat{M}$ be the natural extension of f, and let $\widehat{\mu}$ be the lift of μ, see Section 2.2 for the definitions.

THEOREM 1.8 (Main theorem). *Assume that M, f satisfy assumptions* (A1)–(A7). *For all $\chi > 0$, there is a set* $\mathrm{NUH}^\# \subset \widehat{M}$, *a locally compact countable topological Markov shift* (Σ, σ) *and a Hölder continuous map* $\pi : \Sigma \to \widehat{M}$ *s.t.:*

(1) $\pi \circ \sigma = \widehat{f} \circ \pi$.
(2) $\widehat{\mu}[\mathrm{NUH}^\#] = 1$ *for all f-adapted χ-hyperbolic measures μ.*
(3) $\pi\restriction_{\Sigma^\#} : \Sigma^\# \to \mathrm{NUH}^\#$ *is a surjective finite-to-one map.*

Above, $\Sigma^\#$ is the *recurrent set* of Σ; it carries all σ-invariant probability measures, see Section 2.1 for the definition. In general, π might not be bounded-to-one. In summary, Theorem 1.8 provides a single symbolic extension that is finite-to-one and onto a set that carries all f-adapted χ-hyperbolic measures.

1.2. Method of proof

Let m denote the dimension of M. The proof of Theorem 1.8 can be summarized in the following steps:

(1) The derivative cocycle df induces an invertible cocycle $\widehat{df}$, defined on a fiber bundle over the natural extension space $\widehat{M}$, with the same spectrum as df.
(2) For a fixed $\chi > 0$, define a nonuniformly hyperbolic locus $\mathrm{NUH} \subset \widehat{M}$ of points with nonuniform hyperbolicity at least $\chi > 0$.
(3) For each $\widehat{x} \in \mathrm{NUH}$, define a Pesin chart $\Psi_{\widehat{x}}$ from a subset of $\mathbb{R}^m$ to M for which f and its inverse branches defined by $\widehat{x}$ are small perturbations of hyperbolic matrices.
(4) Define a set $\mathrm{NUH}^* \subset \mathrm{NUH}$ in which there are values $Q(\widehat{x})$ s.t. the restriction of $\Psi_{\widehat{x}}$ to the ball with radius $Q(\widehat{x})$ is well-defined and
$$\lim_{n\to\infty} \tfrac{1}{n} \log Q(\widehat{f}^n(\widehat{x})) = 0.$$
(5) Introduce the ε-double chart $\Psi_{\widehat{x}}^{p^s,p^u}$ as a pair of Pesin charts with the same center but different domains. The parameters p^s/p^u give definite sizes for the stable/unstable manifolds at $\widehat{x}$.
(6) Define abstract counterparts of stable/unstable manifolds in $\widehat{M}$, which we call stable/unstable sets.
(7) Construct a countable collection of ε-double charts that are dense in the space of all ε-double charts, where denseness is defined in terms of finitely many parameters of $\widehat{x}$.
(8) Draw an edge $\Psi_{\widehat{x}}^{p^s,p^u} \to \Psi_{\widehat{y}}^{q^s,q^u}$ if f and its respective inverse branch, when represented in the Pesin charts, are small perturbations of hyperbolic matrices, and if the parameters p^s, q^u are as large as possible.
(9) Using the stable/unstable sets, each path of ε-charts defines an element of $\widehat{M}$, and this coding induces a *locally finite* countable cover on $\widehat{M}$.
(10) Applying an abstract refinement procedure to this cover, the resulting partition defines (Σ, σ) and π that satisfy Theorem 1.8.

In step (9), local finiteness means that each element of the cover intersects finitely many others. It is important to stress the need of this property, since it guarantees that the partition in step (10) is countable. If not, then we could end up with an uncountable partition (e.g., the refinement of the cover of $\mathbb{R}$ by intervals with rational endpoints is uncountable). Local finiteness is essential, and it is the reason that the whole proof is tailored in a way to provide a precise control on all parameters involved in the construction.

The idea of considering double charts is due to Sarig [**Sar13**], and it allows to treat the hyperbolicity in the stable and unstable directions separately. We believe that this is one of the main reasons for the method to work, together with a precise control of all the parameters involved in the construction. The method has been further refined and simplified [**BO18, LS19, LM18, Lim20**]. Since here we allow derivatives to explode, an even finer control of the parameters is required. For instance, while consisting on the general idea that stable/unstable subspaces improve smoothness when backward/forward iterated, the improvement lemma of Section 6.2 needs to be implemented at a scale that is small enough to have a control on derivatives and large enough to code all relevant orbits.

Additionally, neither $\widehat{M}$ nor $\widehat{f}$ are smooth objects. This is a main issue in the proof, specially when dealing with graph transforms. One of the novelties in this paper is to consider invariant sets instead of invariant manifolds and treat them using both a geometrical (their zeroth positions are submanifolds) and an abstract perspective (they are subsets of $\widehat{M}$). This approach is what allows us to construct a Markov partition for the natural extension.

We also mention a novelty in comparison with [**Lim20**]. Even when working with nonuniformly expanding maps, we will use the parameter p^s. Recalling that p^s represents a choice of size for stable manifolds, it has no geometrical meaning in this case, but its use provides a symmetrization between future and past and, more importantly, allows to obtain a locally compact (Σ, σ) in Theorem 1.8. This is better than [**Lim20**], where the graph might have vertices with infinite degree.

1.3. Related literature

Symbolic models for systems with hyperbolicity have a longstanding history that can be traced back at least to the nineteenth century with the study of geodesic flows on surfaces with constant negative curvature by Hadamard [**Had98**]; see also [**KU07**]. During the late sixties and early seventies of the twentieth century, the area saw a great deal of development when symbolic models were constructed for uniformly hyperbolic diffeomorphisms and flows, via the works of Adler and Weiss [**AW67, AW70**], Sinaĭ [**Sin68a, Sin68b**], Bowen [**Bow70, Bow73**], and Ratner [**Rat69, Rat73**]. Since then, other approaches were created, each of them treating different settings. In the sequel we discuss some of them.

Hofbauer towers: Takahashi created a combinatorial method to construct an isomorphism between a large subset X of the natural extension of β-shifts and countable topological Markov shifts [**Tak73**]. Hofbauer showed that X carries all measures of positive entropy and so β-shifts have a unique measure of maximal entropy [**Hof78**]. Later, Hofbauer extended his construction to piecewise continuous interval maps [**Hof79, Hof81**]. The symbolic models obtained by his methods are called *Hofbauer towers*, and they have been extensively used to understand ergodic properties of one-dimensional maps.

Higher dimensional Hofbauer towers: Buzzi constructed Hofbauer towers for piecewise expanding affine maps in any dimension [**Buz97**], for perturbations of fibered products of one-dimensional maps [**Buz98**], and for arbitrary piecewise invertible maps whose entropy generated by the boundary of some dynamically relevant partition is *less* than the topological entropy of the map [**Buz99**]. These towers carry all invariant measures with entropy close enough to the topological entropy of the system.

Inducing schemes: Many systems, although not hyperbolic, do have sets on which it is possible to define a (not necessarily first) return map that is uniformly hyperbolic. This process is known as *inducing*. Hofbauer towers can actually be regarded as inducing schemes to obtain a uniformly expanding map; see [**Bru95**] for this relation. Some invariant measures of the original system lift to the induced one and, when this occurs, the uniform hyperbolicity can be explored to provide ergodic theoretical properties of the original measure. This approach was done for one-dimensional maps [**PS05**], higher dimensional maps that do not have full "boundary entropy" [**PSZ08**], and expanding measures [**Pin11**].

YOCCOZ PUZZLES: Yoccoz constructed Markov structures for quadratic maps of the complex plane, now called *Yoccoz puzzles*, and used them to prove the MLC conjecture for finitely renormalizable parameters; see, e.g., [**Hub93**].

BILLIARDS: Bunimovich, Chernov, and Sinaĭ constructed countable Markov partitions for two dimensional dispersing billiard maps [**BSC90**]. Young constructed an inducing scheme for certain two dimensional dispersing billiard maps and used it to prove exponential decay of correlations [**You98**]. Young's methods are nowadays called *Young towers*, and are known to exist for various billiards and in many other contexts; see, e.g., [**Che99, Mar04**] and references therein. Lima and Matheus constructed countable Markov partitions for two dimensional billiard maps and hyperbolic measures that are adapted to the billiard map [**LM18**].

SYMBOLIC EXTENSIONS: A symbolic extension is an extension by a subshift over a finite alphabet s.t. the projection map is continuous. They reveal a great deal on the entropy structure of topological systems, and are known to exist in various contexts; see [**BD04, DN05, DM09, Bur11, Dow11**] and references therein.

NONUNIFORMLY HYPERBOLIC DIFFEOMORPHISMS AND FLOWS: Katok constructed horseshoes of large topological entropy for $C^{1+\beta}$ diffeomorphisms [**Kat80**]. Such horseshoes usually have zero measure for measures of maximal entropy. Sarig constructed countable Markov partitions with full topological entropy for $C^{1+\beta}$ surface diffeomorphisms [**Sar13**]. Ben Ovadia extended the work of Sarig to any dimension [**BO18**]. Lima and Sarig constructed symbolic models for nonuniformly hyperbolic three dimensional flows with positive speed [**LS19**].

It is important noticing that the projection map π given by Theorem 1.8 is *Hölder continuous*, which is usually not the case for Hofbauer towers neither for symbolic extensions. Also, its restriction to $\Sigma^{\#}$ is *finite-to-one* and hence preserves entropy. Finally, as mentioned above, while Hofbauer towers only work in specific higher dimensional situations, our setting is broad enough to cover various contexts that were not known to have countable Markov partitions neither finite-to-one symbolic extensions by countable Markov shifts.

CHAPTER 2

Preliminaries

2.1. Basic definitions

Let $\mathscr{G} = (V, E)$ be an oriented graph, where V = vertex set and E = edge set. We denote edges by $v \to w$, and we assume that V is countable.

TOPOLOGICAL MARKOV SHIFT (TMS): A *topological Markov shift* (TMS) is a pair (Σ, σ) where

$$\Sigma := \{\mathbb{Z}\text{-indexed paths on } \mathscr{G}\} = \left\{\underline{v} = \{v_n\}_{n\in\mathbb{Z}} \in V^{\mathbb{Z}} : v_n \to v_{n+1}, \forall n \in \mathbb{Z}\right\}$$

and $\sigma : \Sigma \to \Sigma$ is the left shift, $[\sigma(\underline{v})]_n = v_{n+1}$. The *recurrent set* of Σ is

$$\Sigma^\# := \left\{\underline{v} \in \Sigma : \exists v, w \in V \text{ s.t. } \begin{array}{l} v_n = v \text{ for infinitely many } n > 0 \\ v_n = w \text{ for infinitely many } n < 0 \end{array}\right\}.$$

We endow Σ with the distance $d(\underline{v}, \underline{w}) := \exp[-\min\{|n| \in \mathbb{Z} : v_n \neq w_n\}]$. This choice is not canonical, and affects the Hölder regularity of π in Theorem 1.8.

Throughout this text, we will consider subsets of euclidean spaces $\mathbb{R}^n$. All calculations in $\mathbb{R}^n$ will be made with respect to the canonical euclidean norm $\|\cdot\|$ induced by the canonical inner product $\langle\cdot,\cdot\rangle_{\mathbb{R}^n}$ of $\mathbb{R}^n$. Given a linear transformation $T : \mathbb{R}^n \to \mathbb{R}^m$, let $\|T\| := \sup_{\substack{v\in\mathbb{R}^n \\ v\neq 0}} \frac{\|Tv\|}{\|v\|}$. Given an open bounded set $U \subset \mathbb{R}^n$ and $h : U \to \mathbb{R}^m$, let $\|h\|_{C^0} := \sup_{x\in U} \|h(x)\|$ denote the C^0 norm of h. In most cases the domain U will be clear, but when it is not we will write $\|h\|_{C^0(U)}$.

For $0 < \delta \leq 1$, let $\mathrm{Hol}_\delta(h) := \sup \frac{\|h(x)-h(y)\|}{\|x-y\|^\delta}$ where the supremum ranges over distinct elements $x, y \in U$. Observe that $\mathrm{Hol}_1(h)$ is the best Lipschitz constant for h, and so we denote it by $\mathrm{Lip}(h)$. We call h a *contraction* if $\mathrm{Lip}(h) < 1$, and an *expansion* if h is invertible and $\mathrm{Lip}(h^{-1}) < 1$. When h is a linear transformation, then $\|h\| = \mathrm{Lip}(h)$. We also define $\|h\|_{C^\delta} := \|h\|_{C^0} + \mathrm{Hol}_\delta(h)$. Now assume that h is differentiable. By the mean value inequality, $\mathrm{Lip}(h) = \|dh\|_{C^0}$. In this case, define its C^1 norm by $\|h\|_{C^1} := \|h\|_{C^0} + \|dh\|_{C^0} = \|h\|_{C^0} + \mathrm{Lip}(h)$. Finally, for $0 < \delta \leq 1$ let

$$\|h\|_{C^{1+\delta}} := \|h\|_{C^1} + \mathrm{Hol}_\delta(dh) = \|h\|_{C^0} + \|dh\|_{C^\delta} = \|h\|_{C^0} + \|dh\|_{C^0} + \mathrm{Hol}_\delta(dh)$$

denote its $C^{1+\delta}$ norm.

The euclidean norm is equivalent to any other norm in $\mathbb{R}^n$, but the notions of contraction and expansion differ from the choice of the norm. On the other hand, statements claiming that a norm is small are not too sensitive to the choice of the norm. Since it is the euclidean norm that will be related to the definition of the Lyapunov inner product (see Section 3.1), we prefer to work only with the euclidean norm. Given $r > 0$, denote the ball of $\mathbb{R}^n$ with center 0 and radius r by $B^n[r]$. Observe that if $0 \leq d \leq n$ then $B^n[r] \subset B^d[r] \times B^{n-d}[r] \subset B^n[\sqrt{2}r]$.

We fix a smooth Riemannian manifold $(M, \langle\cdot,\cdot\rangle)$ of dimension m and finite diameter. Multiplying the metric by a small constant if necessary, we assume that the diameter of M is smaller than one. We do not require M to be connected neither compact. For $r > 0$, let $B_x[r] \subset T_xM$ denote the open ball with center 0 and radius r. Since M will be fixed throughout the majority of the paper, we will denote the ball of $\mathbb{R}^m$ with center 0 and radius r simply by $B[r]$. Write $a = e^{\pm\varepsilon}b$ when $e^{-\varepsilon} \leq \frac{a}{b} \leq e^{\varepsilon}$, and $a = \pm b$ when $-|b| \leq a \leq |b|$. We use $a \wedge b$ to denote $\min\{a, b\}$.

2.2. Natural extensions

Most of the discussion in this section is classical; see, e.g., [**Roh61**] or [**Aar97**, §3.1]. Given a map $f : M \to M$, let

$$\widehat{M} := \{\widehat{x} = (x_n)_{n\in\mathbb{Z}} : f(x_{n-1}) = x_n, \forall n \in \mathbb{Z}\}.$$

We will write $\widehat{x} = (\ldots, x_{-1}; x_0, x_1, \ldots)$ where ; denotes the separation between the positions -1 and 0. Although $\widehat{M}$ does depend on f, we will not write this dependence. Endow $\widehat{M}$ with the distance $\widehat{d}(\widehat{x}, \widehat{y}) := \sup\{2^n d(x_n, y_n) : n \leq 0\}$; then $\widehat{M}$ is a compact metric space. As for TMS, the definition of $\widehat{d}$ is not canonical and affects the Hölder regularity of π in Theorem 1.8. For each $n \in \mathbb{Z}$, let $\vartheta_n : \widehat{M} \to M$ be the projection into the nth coordinate, $\vartheta_n[\widehat{x}] = x_n$. Consider the sigma-algebra in $\widehat{M}$ generated by $\{\vartheta_n : n \leq 0\}$, i.e. the smallest sigma-algebra that makes all ϑ_n, $n \leq 0$, measurable.

NATURAL EXTENSION OF f: The *natural extension* of f is the map $\widehat{f} : \widehat{M} \to \widehat{M}$ defined by $\widehat{f}(\ldots, x_{-1}; x_0, \ldots) = (\ldots, x_0; f(x_0), \ldots)$. It is an invertible map, with inverse $\widehat{f}^{-1}(\ldots, x_{-1}; x_0, \ldots) = (\ldots, x_{-2}; x_{-1}, \ldots)$.

Note that $\widehat{f}$ is indeed an extension of f, since $\vartheta_0 \circ \widehat{f} = f \circ \vartheta_0$. It is the smallest invertible extension of f: any other invertible extension of f is an extension of $\widehat{f}$. The benefit of considering the natural extension is that, in addition to having an invertible map explicitly defined, there is a bijection between f-invariant and $\widehat{f}$-invariant probability measures, as follows.

PROJECTION OF A MEASURE: If $\widehat{\mu}$ is an $\widehat{f}$-invariant probability measure, then $\mu = \widehat{\mu} \circ \vartheta_0^{-1}$ is an f-invariant probability measure.

LIFT OF A MEASURE: If μ is an f-invariant probability measure, let $\widehat{\mu}$ be the unique probability measure on $\widehat{M}$ s.t. $\widehat{\mu}[\{\widehat{x} \in \widehat{M} : x_n \in A\}] = \mu[A]$ for all $A \subset M$ Borel and all $n \leq 0$.

It is clear that $\widehat{\mu}$ is $\widehat{f}$-invariant. It is also clear that if $\mu, \widehat{\mu}$ are related either by the projection or lift procedures, then for every measurable $\varphi : M \to \mathbb{R}$ it holds $\int_M \varphi d\mu = \int_{\widehat{M}} (\varphi \circ \vartheta_0) d\widehat{\mu}$. What is less clear is that the projection and lift procedures are inverse operations, and that they preserve ergodicity and the Kolmogorov-Sinaĭ entropy, see [**Roh61**]. Here is one consequence of these facts: μ is an ergodic equilibrium measure for a potential $\varphi : M \to \mathbb{R}$ iff $\widehat{\mu}$ is an ergodic equilibrium measure for $\varphi \circ \vartheta_0 : \widehat{M} \to \mathbb{R}$. In particular, the topological entropies of f and $\widehat{f}$ coincide, and μ is a measure of maximal entropy for f iff $\widehat{\mu}$ is a measure of maximal entropy for $\widehat{f}$.

Now let $N = \bigsqcup_{x\in M} N_x$ be a vector bundle over M, and let $A : N \to N$ measurable s.t. for every $x \in M$ the restriction $A \restriction_{N_x}$ is a linear isomorphism $A_x : N_x \to N_{f(x)}$. The map A defines a (possibly non-invertible) cocycle $(A^{(n)})_{n\geq 0}$ over f by $A_x^{(n)} = A_{f^{n-1}(x)} \cdots A_{f(x)} A_x$ for $x \in M$, $n \geq 0$. There is a way of extending $(A^{(n)})_{n\geq 0}$ to an invertible cocycle over $\widehat{f}$. For $\widehat{x} \in \widehat{M}$, let $N_{\widehat{x}} := N_{\vartheta_0[\widehat{x}]}$ and let $\widehat{N} := \bigsqcup_{\widehat{x}\in\widehat{M}} N_{\widehat{x}}$, a vector bundle over $\widehat{M}$. Define the map $\widehat{A} : \widehat{N} \to \widehat{N}$, $\widehat{A}_{\widehat{x}} := A_{\vartheta_0[\widehat{x}]}$. For $\widehat{x} = (x_n)_{n\in\mathbb{Z}}$, define

$$\widehat{A}_{\widehat{x}}^{(n)} := \begin{cases} A_{x_0}^{(n)} & \text{, if } n \geq 0 \\ A_{x_{-n}}^{-1} \cdots A_{x_{-2}}^{-1} A_{x_{-1}}^{-1} & \text{, if } n \leq 0. \end{cases}$$

By definition, $\widehat{A}_{\widehat{x}}^{(m+n)} = \widehat{A}_{\widehat{f}^n(\widehat{x})}^{(m)} \widehat{A}_{\widehat{x}}^{(n)}$ for all $\widehat{x} \in \widehat{M}$ and all $m, n \in \mathbb{Z}$, hence $(\widehat{A}^{(n)})_{n\in\mathbb{Z}}$ is an invertible cocycle over $\widehat{f}$.

Let M, f satisfying (A1)–(A7), and take $N = TM$ and $A = df$. Define $\rho : \widehat{M} \to \mathbb{R}$ by $\rho(\widehat{x}) := d(\{\vartheta_{-1}[\widehat{x}], \vartheta_0[\widehat{x}], \vartheta_1[\widehat{x}]\}, \mathscr{S})$. We note that μ is f-adapted iff $\log \rho \in L^1(\widehat{\mu})$:

- If μ is f-adapted then by $\widehat{f}$-invariance the functions $\log d(\vartheta_i[\widehat{x}], \mathscr{S})$, $i = -1, 0, 1$, are in $L^1(\widehat{\mu})$ and so is their minimum $\log \rho$.
- Reversely, if $\log \rho \in L^1(\widehat{\mu})$ then $\log d(\vartheta_0[\widehat{x}], \mathscr{S}) \in L^1(\widehat{\mu})$ and so

$$\int_M |\log d(x, \mathscr{S})| d\mu(x) = \int_{\widehat{M}} |\log d(\vartheta_0[\widehat{x}], \mathscr{S})| d\widehat{\mu}(\widehat{x}) < \infty,$$

thus proving that $\log d(x, \mathscr{S}) \in L^1(\mu)$.

For simplicity, we will write ϑ to represent ϑ_0. If θ is a function on M, we will sometimes represent the composition $\theta \circ \vartheta$ also by θ. For example, $\mathfrak{d}(\widehat{x}) := \mathfrak{d}(\vartheta[\widehat{x}])$ and $\mathfrak{B}_{\widehat{x}} := \mathfrak{B}_{\vartheta[\widehat{x}]}$.

Part 1

Proof of the Main Theorem

CHAPTER 3

Pesin theory

Consider the derivative cocycle $(df_x^n)_{n\geq 0}$. Since df_x is not necessarily invertible for all $x\in M$, we apply the method described in the previous section to extend it to an invertible cocycle, defined in a subset of $\widehat{M}$.

INVERTIBLE COCYCLE: The cocycle $(df_x^n)_{n\geq 0}$ induces an invertible cocycle $(\widehat{df}_{\widehat{x}}^{(n)})_{n\in\mathbb{Z}}$, defined for all $\widehat{x}=(x_n)_{n\in\mathbb{Z}}\in\widehat{M}\backslash\bigcup_{n\in\mathbb{Z}}\widehat{f}^n(\vartheta^{-1}[\mathscr{S}])$ by

$$\widehat{df}_{\widehat{x}}^{(n)}:=\begin{cases} df_{x_0}^n & \text{, if } n\geq 0\\ (df_{x_{-n}})^{-1}\cdots(df_{x_{-2}})^{-1}(df_{x_{-1}})^{-1} & \text{, if } n\leq 0.\end{cases}$$

Since $\widehat{x}\notin\bigcup_{n\in\mathbb{Z}}\widehat{f}^n(\vartheta^{-1}[\mathscr{S}])$, each df_{x_n} is invertible and so the above definition makes sense. The vector bundle in which $(\widehat{df}_{\widehat{x}}^{(n)})_{n\in\mathbb{Z}}$ acts is $\widehat{TM}$, where $\widehat{TM}_{\widehat{x}}=T_{x_0}M$. We consider a parameter $\chi>0$, fixed throughout all of Part 1.

THE NONUNIFORMLY HYPERBOLIC LOCUS NUH: It is defined as the set of points $\widehat{x}\in\widehat{M}\backslash\bigcup_{n\in\mathbb{Z}}\widehat{f}^n(\vartheta^{-1}[\mathscr{S}])$ for which there is a splitting $\widehat{TM}_{\widehat{x}}=E^s_{\widehat{x}}\oplus E^u_{\widehat{x}}$ s.t.:

(NUH1) Every $v\in E^s_{\widehat{x}}$ contracts in the future at least $-\chi$ and expands in the past:

$$\limsup_{n\to+\infty}\tfrac{1}{n}\log\|\widehat{df}^{(n)}v\|\leq-\chi \quad\text{and}\quad \liminf_{n\to+\infty}\tfrac{1}{n}\log\|\widehat{df}^{(-n)}v\|>0.$$

(NUH2) Every $v\in E^u_{\widehat{x}}$ contracts in the past at least $-\chi$ and expands in the future:

$$\limsup_{n\to+\infty}\tfrac{1}{n}\log\|\widehat{df}^{(-n)}v\|\leq-\chi \text{ and } \liminf_{n\to+\infty}\tfrac{1}{n}\log\|\widehat{df}^{(n)}v\|>0.$$

(NUH3) The parameters $s(\widehat{x})=\sup\limits_{\substack{v\in E^s_{\widehat{x}}\\ \|v\|=1}}S(\widehat{x},v)$ and $u(\widehat{x})=\sup\limits_{\substack{w\in E^u_{\widehat{x}}\\ \|w\|=1}}U(\widehat{x},w)$ are finite, where:

$$S(\widehat{x},v)=\sqrt{2}\left(\sum_{n\geq 0}e^{2n\chi}\|\widehat{df}^{(n)}v\|^2\right)^{1/2},$$

$$U(\widehat{x},w)=\sqrt{2}\left(\sum_{n\geq 0}e^{2n\chi}\|\widehat{df}^{(-n)}w\|^2\right)^{1/2}.$$

The above conditions are weaker than Lyapunov regularity, hence NUH contains all Lyapunov regular points with exponents greater than χ in absolute value. Moreover, a periodic point $\widehat{x}$ is in NUH iff all of its exponents are greater than χ in absolute value. But NUH might contain points with some Lyapunov exponents equal to $\pm\chi$, and even non-regular points, where the contraction rates oscillate infinitely often. Formally, $\text{NUH}=\text{NUH}_\chi$ depends on $\chi>0$, but since this parameter is fixed, we will omit such dependence. When we need to consider NUH for different values of χ, we will be explicit. Clearly, NUH is $\widehat{f}$-invariant. By the Oseledets theorem, if μ is χ-hyperbolic then (NUH1)–(NUH3) hold $\widehat{\mu}$-a.e.

3.1. Oseledets–Pesin reduction

We provide a diagonalization for the cocycle $(\widehat{df}^{(n)}_{\widehat{x}})_{n\in\mathbb{Z}}$ inside NUH. Write $E^s \oplus E^u$ to represent the splitting on NUH.

LYAPUNOV INNER PRODUCT: We define an inner product $\langle\!\langle \cdot, \cdot \rangle\!\rangle$ on NUH, which we call *Lyapunov inner product*, by the following identities:

- For $v_1^s, v_2^s \in E^s$:

$$\langle\!\langle v_1^s, v_2^s \rangle\!\rangle = 2\sum_{n\geq 0} e^{2n\chi} \left\langle \widehat{df}^{(n)} v_1^s, \widehat{df}^{(n)} v_2^s \right\rangle.$$

- For $v_1^u, v_2^u \in E^u$:

$$\langle\!\langle v_1^u, v_2^u \rangle\!\rangle = 2\sum_{n\geq 0} e^{2n\chi} \left\langle \widehat{df}^{(-n)} v_1^u, \widehat{df}^{(-n)} v_2^u \right\rangle.$$

- For $v^s \in E^s$ and $v^u \in E^u$:

$$\langle\!\langle v^s, v^u \rangle\!\rangle = 0.$$

By conditions (NUH1)–(NUH3), the infinite sums above are finite. Let $|\!|\!|\cdot|\!|\!|$ denote the norm induced by $\langle\!\langle \cdot, \cdot \rangle\!\rangle$. As we know, $|\!|\!|\cdot|\!|\!|$ uniquely defines $\langle\!\langle \cdot, \cdot \rangle\!\rangle$, so we also call $|\!|\!|\cdot|\!|\!|$ the Lyapunov inner product. Because of the term $n = 0$ that appears in the sums defining $\langle\!\langle \cdot, \cdot \rangle\!\rangle$, we have $|\!|\!|v|\!|\!| \geq \sqrt{2}\|v\|$ for all $v \in E^s \oplus E^u$. Furthermore, if $v^s \in E^s_{\widehat{x}}$ then $|\!|\!|v^s|\!|\!| = S(\widehat{x}, v^s)$, and if $v^u \in E^u_{\widehat{x}}$ then $|\!|\!|v^u|\!|\!| = U(\widehat{x}, v^u)$.

For $\widehat{x} \in$ NUH, let $d_s(\widehat{x}), d_u(\widehat{x}) \in \mathbb{N}$ be the dimensions of $E^s_{\widehat{x}}, E^u_{\widehat{x}}$ respectively. Since the splitting $E^s \oplus E^u$ is $\widehat{df}$-invariant, the functions d_s, d_u are $\widehat{f}$-invariant. Recall that $\langle \cdot, \cdot \rangle_{\mathbb{R}^n}$ denotes the canonical inner product in $\mathbb{R}^n$.

LINEAR MAP $C(\widehat{x})$: For $\widehat{x} \in$ NUH, define $C(\widehat{x}) : \mathbb{R}^n \to \widehat{TM}_{\widehat{x}}$ to be a linear map satisfying the following conditions:

- $C(\widehat{x})$ sends the subspace $\mathbb{R}^{d_s(\widehat{x})} \times \{0\}$ to $E^s_{\widehat{x}}$ and $\{0\} \times \mathbb{R}^{d_u(\widehat{x})}$ to $E^u_{\widehat{x}}$.
- $\langle v, w \rangle_{\mathbb{R}^n} = \langle\!\langle C(\widehat{x})v, C(\widehat{x})w \rangle\!\rangle$ for all $v, w \in \mathbb{R}^n$, i.e., $C(\widehat{x})$ is an isometry between $(\mathbb{R}^n, \langle \cdot, \cdot \rangle_{\mathbb{R}^n})$ and $(\widehat{TM}_{\widehat{x}}, \langle\!\langle \cdot, \cdot \rangle\!\rangle)$.

The map $C(\widehat{x})$ is not uniquely defined (for instance, rotations inside $E^s_{\widehat{x}}, E^u_{\widehat{x}}$ preserve both properties above). What is important for us is to define $C(\widehat{x})$ s.t. the map $\widehat{x} \in \text{NUH} \mapsto C(\widehat{x})$ is measurable, and this can be done as in [**BO18**, Footnote at page 48]. Below we list the main properties of $C(\widehat{x})$.

LEMMA 3.1. *For all $\widehat{x} \in$ NUH, the following holds.*

(1) $\|C(\widehat{x})\| \leq 1$ *and*

$$\|C(\widehat{x})^{-1}\|^2 = \sup_{\substack{v^s + v^u \in E^s_{\widehat{x}} \oplus E^u_{\widehat{x}} \\ \|v^s + v^u\| \neq 0}} \frac{|\!|\!|v^s|\!|\!|^2 + |\!|\!|v^u|\!|\!|^2}{\|v^s + v^u\|^2}.$$

(2) $D(\widehat{x}) = C(\widehat{f}(\widehat{x}))^{-1} \circ \widehat{df}_{\widehat{x}} \circ C(\widehat{x})$ *is a block matrix*

$$D(\widehat{x}) = \begin{bmatrix} D_s(\widehat{x}) & 0 \\ 0 & D_u(\widehat{x}) \end{bmatrix}$$

where $D_s(\widehat{x})$ *is a* $d_s(\widehat{x}) \times d_s(\widehat{x})$ *matrix with* $\|D_s(\widehat{x})v_1\| < e^{-\chi}$ *for all unit vectors* $v_1 \in \mathbb{R}^{d_s(\widehat{x})}$, *and* $D_u(\widehat{x})$ *is a* $d_u(\widehat{x}) \times d_u(\widehat{x})$ *matrix with* $\|D_u(\widehat{x})v_2\| > e^{\chi}$ *for all unit vectors* $v_2 \in \mathbb{R}^{d_u(\widehat{x})}$.

(3) *If* $v_1 \in \mathbb{R}^{d_s(\widehat{x})}$ *is unitary then*

$$\|D_s(\widehat{x})v_1\| \geq d(\vartheta[\widehat{x}], \mathscr{S})^a (1 + e^{2\chi})^{-1/2},$$

and if $v_2 \in \mathbb{R}^{d_u(\widehat{x})}$ *is unitary then*

$$\|D_u(\widehat{x})v_2\| \leq d(\vartheta[\widehat{x}], \mathscr{S})^{-a} (1 + e^{2\chi})^{1/2}.$$

(4) $d(\vartheta[\widehat{x}], \mathscr{S})^{2a}(1 + e^{2\chi})^{-1/2} \leq \frac{\|C(\widehat{f}(\widehat{x}))^{-1}\|}{\|C(\widehat{x})^{-1}\|} \leq d(\vartheta[\widehat{x}], \mathscr{S})^{-2a}(1 + e^{2\chi})^{1/2}$.

PROOF. Parts (1) and (2) can be proved as in [**BO18**, Lemma 2.9], but for later use on some estimates we will prove them. Let $v = v_1 + v_2$, where $v_1 \in \mathbb{R}^{d_s(\widehat{x})} \times \{0\}$ and $v_2 \in \{0\} \times \mathbb{R}^{d_u(\widehat{x})}$. By the Cauchy-Schwarz inequality and the arithmetic-geometric mean inequality, we have

$$2\langle C(\widehat{x})v_1, C(\widehat{x})v_2 \rangle \leq 2\|C(\widehat{x})v_1\| \cdot \|C(\widehat{x})v_2\| \leq \|C(\widehat{x})v_1\|^2 + \|C(\widehat{x})v_2\|^2,$$

hence

$$\begin{aligned} \|C(\widehat{x})v\|^2 &= \|C(\widehat{x})v_1\|^2 + \|C(\widehat{x})v_2\|^2 + 2\langle C(\widehat{x})v_1, C(\widehat{x})v_2 \rangle \\ &\leq 2(\|C(\widehat{x})v_1\|^2 + \|C(\widehat{x})v_2\|^2) \leq |||C(\widehat{x})v_1|||^2 + |||C(\widehat{x})v_2|||^2 \\ &= \|v_1\|^2 + \|v_2\|^2 \\ &= \|v\|^2, \end{aligned}$$

thus proving that $\|C(\widehat{x})\| \leq 1$. The formula for $\|C(\widehat{x})^{-1}\|$ is almost automatic: if $v = v^s + v^u \in E^s_{\widehat{x}} \oplus E^u_{\widehat{x}}$ then $\|C(\widehat{x})^{-1}v\|^2 = \langle\langle v, v \rangle\rangle = |||v^s|||^2 + |||v^u|||^2$ and so

$$\|C(\widehat{x})^{-1}\|^2 = \sup_{\substack{v \in \widehat{TM}_{\widehat{x}} \\ v \neq 0}} \frac{|||v^s|||^2 + |||v^u|||^2}{\|v\|^2}.$$

This proves part (1). Now we prove part (2). By the definition of $C(\cdot)$, we have

$$\mathbb{R}^{d_s(\widehat{x})} \times \{0\} \xrightarrow{C(\widehat{x})} E^s_{\widehat{x}} \xrightarrow{\widehat{df}_{\widehat{x}}} E^s_{\widehat{f}(\widehat{x})} \xrightarrow{C(\widehat{f}(\widehat{x}))^{-1}} \mathbb{R}^{d_s(\widehat{f}(\widehat{x}))} \times \{0\}.$$

Similarly,

$$\{0\} \times \mathbb{R}^{d_u(\widehat{x})} \xrightarrow{C(\widehat{x})} E^u_{\widehat{x}} \xrightarrow{\widehat{df}_{\widehat{x}}} E^u_{\widehat{f}(\widehat{x})} \xrightarrow{C(\widehat{f}(\widehat{x}))^{-1}} \{0\} \times \mathbb{R}^{d_u(\widehat{f}(\widehat{x}))},$$

and so $D(\widehat{x})$ has the block form

$$D(\widehat{x}) = \begin{bmatrix} D_s(\widehat{x}) & 0 \\ 0 & D_u(\widehat{x}) \end{bmatrix},$$

where $D_s(\widehat{x})$ has dimension $d_s(\widehat{x})$ and $D_u(\widehat{x})$ has dimension $d_u(\widehat{x})$. To obtain the estimates of $\|D_s(\widehat{x})\|$ and $\|D_u(\widehat{x})\|$, we give general estimates for the ratios

$$\frac{|||\widehat{df}v^s|||^2}{|||v^s|||^2} \quad \text{and} \quad \frac{|||\widehat{df}v^u|||^2}{|||v^u|||^2}.$$

CLAIM: For all nonzero $v^s \in E^s, v^u \in E^u$ it holds

$$d(\vartheta[\widehat{x}],\mathscr{S})^a(1+e^{2\chi})^{-1/2} < \frac{|||\widehat{df}v^s|||}{|||v^s|||} < e^{-\chi} \text{ and}$$
$$e^{\chi} < \frac{|||\widehat{df}v^u|||}{|||v^u|||} < d(\vartheta[\widehat{x}],\mathscr{S})^{-a}(1+e^{2\chi})^{1/2}.$$

PROOF OF THE CLAIM. Write $\widehat{x}=(x_n)_{n\in\mathbb{Z}}$. Observe that for $v^s\in E^s$ we have

$$|||v^s|||^2 = 2\|v^s\|^2 + 2e^{2\chi}\sum_{n\geq 0} e^{2n\chi}\|\widehat{df}^{(n)}(\widehat{df}v^s)\|^2 = 2\|v^s\|^2 + e^{2\chi}\left|\left|\left|\widehat{df}v^s\right|\right|\right|^2 \tag{3.1}$$

and similarly for $v^u\in E^u$ we have

$$\begin{aligned}|||v^u|||^2 &= 2\|v^u\|^2 + 2e^{2\chi}\sum_{n\geq 0} e^{2n\chi}\left\|\widehat{df}^{(-n)}(\widehat{df}^{(-1)}v^u)\right\|^2 \\ &= 2\|v^u\|^2 + e^{2\chi}\left|\left|\left|\widehat{df}^{(-1)}v^u\right|\right|\right|^2.\end{aligned} \tag{3.2}$$

Equation (3.1) implies that

$$\frac{|||\widehat{df}v^s|||^2}{|||v^s|||^2} = e^{-2\chi}\left(1-\frac{2\|v^s\|^2}{|||v^s|||^2}\right).$$

On one hand, $\frac{|||\widehat{df}v^s|||^2}{|||v^s|||^2} < e^{-2\chi}$. On the other hand, since $|||v^s|||^2 > 2\|v^s\|^2 + 2e^{2\chi}\|\widehat{df}v^s\|^2 \geq 2[1+e^{2\chi}d(x_0,\mathscr{S})^{2a}]\|v^s\|^2$, we have $\frac{2\|v^s\|^2}{|||v^s|||^2} < \frac{1}{1+e^{2\chi}d(x_0,\mathscr{S})^{2a}}$ and so

$$\frac{|||\widehat{df}v^s|||^2}{|||v^s|||^2} > e^{-2\chi}\left(1-\frac{1}{1+e^{2\chi}d(x_0,\mathscr{S})^{2a}}\right) = \frac{d(x_0,\mathscr{S})^{2a}}{1+e^{2\chi}d(x_0,\mathscr{S})^{2a}} > \frac{d(x_0,\mathscr{S})^{2a}}{1+e^{2\chi}}.$$

Hence $d(x_0,\mathscr{S})^a(1+e^{2\chi})^{-1/2} < \frac{|||\widehat{df}v^s|||}{|||v^s|||} < e^{-\chi}$.

The second estimate is proved similarly. Using equation (3.2) with $\widehat{df}v^u$ instead of v^u, we have $\left|\left|\left|\widehat{df}v^u\right|\right|\right|^2 = 2\|\widehat{df}v^u\|^2 + e^{2\chi}|||v^u|||^2$ and so

$$\frac{|||v^u|||^2}{|||\widehat{df}v^u|||^2} = e^{-2\chi}\left(1-\frac{2\|\widehat{df}v^u\|^2}{|||\widehat{df}v^u|||^2}\right).$$

Then $\frac{|||v^u|||^2}{|||\widehat{df}v^u|||^2} < e^{-2\chi}$ and, since $\left|\left|\left|\widehat{df}v^u\right|\right|\right|^2 > 2\|\widehat{df}v^u\|^2 + 2e^{2\chi}\|\widehat{df}^{(-1)}(\widehat{df}v^u)\|^2 \geq 2[1+e^{2\chi}d(x_0,\mathscr{S})^{2a}]\|\widehat{df}v^u\|^2$, we also have

$$\frac{|||v^u|||^2}{|||\widehat{df}v^u|||^2} > e^{-2\chi}\left(1-\frac{1}{1+e^{2\chi}d(x_0,\mathscr{S})^{2a}}\right) > \frac{d(x_0,\mathscr{S})^{2a}}{1+e^{2\chi}},$$

thus implying that $e^{\chi} < \frac{|||\widehat{df}v^u|||}{|||v^u|||} < d(x_0,\mathscr{S})^{-a}(1+e^{2\chi})^{1/2}$. □

The claim actually implies parts (2) and (3), since if $v^s\in E^s$ and $v^u\in E^u$ are nonzero vectors then the definition of $C(\widehat{x})$ implies that

$$\frac{\|D_s(\widehat{x})v^s\|}{\|v^s\|} = \frac{\left|\left|\left|\widehat{df}\circ C(\widehat{x})v^s\right|\right|\right|}{|||C(\widehat{x})v^s|||} \text{ and } \frac{\|D_u(\widehat{x})v^u\|}{\|v^u\|} = \frac{\left|\left|\left|\widehat{df}\circ C(\widehat{x})v^u\right|\right|\right|}{|||C(\widehat{x})v^u|||}.$$

To prove part (4), observe that since $\widehat{df}_{\widehat{x}}$ is invertible, part (1) implies that

$$\left\|C(\widehat{f}(\widehat{x}))^{-1}\right\|^2 = \sup_{\substack{v^s+v^u\in E^s_{\widehat{x}}\oplus E^u_{\widehat{x}}\\ \|v^s+v^u\|\neq 0}} \frac{|||\widehat{df}v^s|||^2+|||\widehat{df}v^u|||^2}{\left\|\widehat{df}(v^s+v^u)\right\|^2}.$$

Now apply the claim and (A6) to conclude part (4). □

In summary, what we did above was to define, inside NUH, a Lyapunov inner product that depends on $\chi > 0$. The construction works because of conditions (NUH1)–(NUH3). For $\chi' < \chi$, we can perform the same construction inside $\text{NUH}_{\chi'}$, obtaining a different Lyapunov inner product $|||\cdot|||_{\chi'}$. Since $\text{NUH}_{\chi'} \supset \text{NUH}$, the Lyapunov inner product $|||\cdot|||_{\chi'}$ is also defined inside NUH. Notice that $|||\cdot|||_{\chi'} \leq |||\cdot|||$, so we say that $|||\cdot|||$ is stronger than $|||\cdot|||_{\chi'}$. For $\widehat{x} \in \text{NUH}_{\chi'}$, define a linear map $C_{\chi'}(\widehat{x})$ using $|||\cdot|||_{\chi'}$. Then $C_{\chi'}(\widehat{x})$ satisfies an analogue of Lemma 3.1 for a block matrix $D_{\chi'}(\widehat{x})$, with χ replaced by χ'. Note that $D(\widehat{x})$ has better hyperbolicity rates than $D_{\chi'}(\widehat{x})$. Also, since $|||\cdot|||_{\chi'} \leq |||\cdot|||$ then part (1) implies that $\left\|C_{\chi'}(\widehat{x})^{-1}\right\| \leq \left\|C(\widehat{x})^{-1}\right\|$.

The Lyapunov inner product $|||\cdot|||$ will be used in the majority of the paper, but there is one subtle point that we will need to use $|||\cdot|||_{\chi'}$ (see Section 6.3), as successfully implemented by Ben Ovadia [**BO21**].

3.2. Pesin charts

The next step is to define charts for which $\widehat{f}$ is a small perturbation of a hyperbolic matrix. Recall that $\mathfrak{d}(\widehat{x}) := \mathfrak{d}(\vartheta[\widehat{x}])$.

PESIN CHART $\Psi_{\widehat{x}}$: The *Pesin chart* at $\widehat{x} \in \text{NUH}$ is $\Psi_{\widehat{x}} : B[\mathfrak{d}(\widehat{x})] \to M$ defined by $\Psi_{\widehat{x}} := \exp_{\vartheta[\widehat{x}]} \circ C(\widehat{x})$.

LEMMA 3.2. *The Pesin chart $\Psi_{\widehat{x}}$ is a diffeomorphism onto its image s.t.:*

(1) $\text{Lip}\,(\Psi_{\widehat{x}}) \leq 2$ *and* $\text{Lip}\left(\Psi_{\widehat{x}}^{-1}\right) \leq 2\|C(\widehat{x})^{-1}\|$.

(2) $\|\widetilde{d(\Psi_{\widehat{x}})_{v_1}} - \widetilde{d(\Psi_{\widehat{x}})_{v_2}}\| \leq d(\vartheta[\widehat{x}], \mathscr{S})^{-a}\|v_1 - v_2\|$ *for all* $v_1, v_2 \in B[\mathfrak{d}(\widehat{x})]$.

PROOF. The proof is very similar to the proof of [**LM18**, Lemma 3.1]. Write $\widehat{x} = (x_n)_{n\in\mathbb{Z}}$. Since $C(\widehat{x})$ is a contraction, we have $C(\widehat{x})(B[\mathfrak{d}(x_0)]) \subset B_{x_0}[\mathfrak{d}(x_0)]$ and so by (A1) it follows that $\Psi_{\widehat{x}}$ is a well-defined differentiable map with $\Psi_{\widehat{x}}(B[\mathfrak{d}(x_0)]) \subset \mathfrak{B}_{x_0}$. Its inverse is $C(\widehat{x})^{-1} \circ \exp_{x_0}^{-1}$ which, again by (A1), is a well-defined differentiable map from $\Psi_{\widehat{x}}(B[\mathfrak{d}(x_0)])$ onto $B[\mathfrak{d}(x_0)]$. This proves that $\Psi_{\widehat{x}}$ is a diffeomorphism onto its image.

(1) By Lemma 3.1(1) and (A2), $\text{Lip}\,(\Psi_{\widehat{x}}) \leq 2$ and $\text{Lip}\left(\Psi_{\widehat{x}}^{-1}\right) \leq 2\|C(\widehat{x})^{-1}\|$.

(2) Since $\|C(\widehat{x})v_i\| \leq \|v_i\| \leq \mathfrak{d}(x_0)$, assumption (A3) implies that

$$\begin{aligned}
\|\widetilde{d(\Psi_{\widehat{x}})_{v_1}} - \widetilde{d(\Psi_{\widehat{x}})_{v_2}}\| &= \|\widetilde{d(\exp_{x_0})_{C(\widehat{x})v_1}} \circ C(\widehat{x}) - \widetilde{d(\exp_{x_0})_{C(\widehat{x})v_2}} \circ C(\widehat{x})\| \\
&\leq d(x_0, \mathscr{S})^{-a}\|C(\widehat{x})v_1 - C(\widehat{x})v_2\| \\
&\leq d(x_0, \mathscr{S})^{-a}\|v_1 - v_2\|,
\end{aligned}$$

thus completing the proof. □

Now we introduce a parameter for which the restriction of $\Psi_{\widehat{x}}$ to a smaller domain with this size has better properties. Given $\varepsilon > 0$, let $I_\varepsilon := \{e^{-\frac{1}{3}\varepsilon n} : n \geq 0\}$. Recall that $\rho(\widehat{x}) = d(\{\vartheta_{-1}[\widehat{x}], \vartheta_0[\widehat{x}], \vartheta_1[\widehat{x}]\}, \mathscr{S})$.

PARAMETER $Q(\widehat{x})$: For $\widehat{x} \in \text{NUH}$, define $Q(\widehat{x}) = \max\{Q \in I_\varepsilon : Q \leq \widetilde{Q}(\widehat{x})\}$, where

$$\widetilde{Q}(\widehat{x}) = \varepsilon^{6/\beta} \min\left\{\left\|C(\widehat{x})^{-1}\right\|^{-48/\beta}, \rho(\widehat{x})^{96a/\beta}\right\}.$$

This parameter depends on $\varepsilon > 0$, but for simplicity we will suppress this dependence. It is important noting that most of the results will hold for $\varepsilon > 0$ small enough. The term $\varepsilon^{6/\beta}$ will allow to absorb multiplicative constants, while the minimum takes into account the two situations where we will have to take a smaller domain for the Pesin chart:

- The rate of nonuniform hyperbolicity is measured by $\|C(\widehat{x})^{-1}\|$: the longer it takes to see hyperbolicity, the bigger $\|C(\widehat{x})^{-1}\|$ is.
- The term $\rho(\widehat{x})$ controls how small the Pesin charts needs to be for its image not to intersect $\mathscr{S}$.

We observe that this definition is stronger than those in [**Sar13, LS19, LM18**]. In particular, calculations in these works that only depend on the definition of $Q(\widehat{x})$ can be used here. From its definition and Lemma 3.1(4), we have the following bounds for $Q(\widehat{x})$ for $\varepsilon > 0$ small enough:

$$
\begin{aligned}
&Q(\widehat{x}) \leq \varepsilon^{6/\beta}, \ \|C(\widehat{x})^{-1}\| Q(\widehat{x})^{\beta/48} \leq \varepsilon^{1/8}, \\
&\|C(\widehat{f}(\widehat{x}))^{-1}\| Q(\widehat{x})^{\beta/12} \leq \varepsilon^{1/4}, \ \rho(\widehat{x})^{-a} Q(\widehat{x})^{\beta/96} < \varepsilon^{1/16}.
\end{aligned} \tag{3.3}
$$

We can also define, for each $\widehat{x} \in \mathrm{NUH}_{\chi'}$, a Pesin chart $\Psi_{\chi',\widehat{x}} = \exp_{\vartheta[\widehat{x}]} \circ C_{\chi'}(\widehat{x})$, for which Lemma 3.2 holds with χ replaced by χ'. We could also define a parameter $Q_{\chi'}(\widehat{x})$, where $\|C(\widehat{x})^{-1}\|$ is replaced by $\|C_{\chi'}(\widehat{x})^{-1}\|$. In this case, for $\widehat{x} \in \mathrm{NUH}$ we have $Q(\widehat{x}) \leq Q_{\chi'}(\widehat{x})$. Since all calculations will be made with the smaller value $Q(\widehat{x})$, the parameter $Q_{\chi'}$ has no essential role in this paper.

3.3. The map f and its inverse branches in Pesin charts

We start introducing a notation for the inverse branches of f.

Inverse branches of f: If $x, f(x) \notin \mathscr{S}$, let f_x^{-1} be the inverse branch of f that sends $f(x)$ to x and is defined in $E_x = B(f(x), 2\mathfrak{r}(x))$. Given $\widehat{x} = (x_n)_{n\in\mathbb{Z}} \in \widehat{M}$, write $f_{\widehat{x}}^{-1} := f_{x_0}^{-1}$.

By assumption (A5), f_x is a well-defined diffeomorphism from E_x onto its image. For $x \in M$, define $\mathfrak{m}(x) := \frac{1}{2}\min\{d(x,\mathscr{S})^{2a}, d(f(x),\mathscr{S})^{2a}\}$, and $\mathfrak{m}(\widehat{x}) := \mathfrak{m}(\vartheta[\widehat{x}])$.

The maps $F_{\widehat{x}}$ and $F_{\widehat{x}}^{-1}$: For $\widehat{x} \in \mathrm{NUH}$, define $F_{\widehat{x}} : B[\mathfrak{m}(\widehat{x})] \to \mathbb{R}^m$ by the composition $F_{\widehat{x}} = \Psi_{\widehat{f}(\widehat{x})}^{-1} \circ f \circ \Psi_{\widehat{x}}$; similarly, define $F_{\widehat{x}}^{-1} : B[\mathfrak{m}(\widehat{x})] \to \mathbb{R}^m$ by $F_{\widehat{x}}^{-1} = \Psi_{\widehat{x}}^{-1} \circ f_{\widehat{x}}^{-1} \circ \Psi_{\widehat{f}(\widehat{x})}$.

Theorem 3.3. *The following holds for all $\varepsilon > 0$ small enough. If $\widehat{x} \in \mathrm{NUH}$ then $F_{\widehat{x}}, F_{\widehat{x}}^{-1}$ are well-defined diffeomorphisms onto their images. Furthermore, the restrictions of $F_{\widehat{x}}$ and $F_{\widehat{x}}^{-1}$ to $B[20Q(\widehat{x})]$ and $B[20Q(\widehat{f}(\widehat{x}))]$ respectively satisfy:*

1. *$F_{\widehat{x}}^{-1} \circ F_{\widehat{x}}$ and $F_{\widehat{x}} \circ F_{\widehat{x}}^{-1}$ are the identity map.*
2. *$d(F_{\widehat{x}}^{\pm 1})_0 = D(\widehat{x})^{\pm 1}$, cf. Lemma 3.1.*
3. *$F_{\widehat{x}} = D(\widehat{x}) + H^+$ and $F_{\widehat{x}}^{-1} = D(\widehat{x})^{-1} + H^-$ where:*
   (a) *$H^{\pm}(0) = 0$ and $d(H^{\pm})_0 = 0$.*
   (b) *$\|H^{\pm}\|_{C^{1+\frac{\beta}{2}}} < \varepsilon$.*
4. *$\|dF_{\widehat{x}}^{\pm 1}\|_{C^0} < 2d(\vartheta[\widehat{x}], \mathscr{S})^{-a}(1 + e^{2\chi})^{1/2}$.*

PROOF. The proof is an adaptation of the proof of [**LM18**, Theorem 3.3] to our context. We start showing that $F_{\widehat{x}}^{\pm 1}$ are well-defined. Write $\widehat{x} = (x_n)_{n\in\mathbb{Z}}$. Given $v \in B[\mathfrak{m}(\widehat{x})]$, Lemma 3.2(1) implies that $\Psi_{\widehat{x}}(v) \in B(x_0, 2\mathfrak{m}(\widehat{x})) \subset D_{x_0}$ and by (A6) we get that $(f \circ \Psi_{\widehat{x}})(v) \in B(f(x_0), 2\mathfrak{m}(\widehat{x})d(x_0,\mathscr{S})^{-a}) \subset \mathfrak{B}_{f(x_0)}$, where in the latter inclusion we used that $2\mathfrak{m}(\widehat{x})d(x_0,\mathscr{S})^{-a} \leq d(f(x_0),\mathscr{S})^a < \mathfrak{d}(f(x_0))$. By (A1), the composition $(\Psi_{\widehat{f(x)}}^{-1} \circ f \circ \Psi_{\widehat{x}})(v)$ is well-defined. We proceed similarly for $F_{\widehat{x}}^{-1}$: given $v \in B[\mathfrak{m}(\widehat{x})]$, we have $\Psi_{\widehat{f(x)}}(v) \in B(f(x_0), 2\mathfrak{m}(\widehat{x})) \subset E_{x_0}$ and so (A6) gives that $(f_{\widehat{x}}^{-1} \circ \Psi_{\widehat{f(x)}})(v) \in B(x_0, 2\mathfrak{m}(\widehat{x})d(x_0,\mathscr{S})^{-a}) \subset \mathfrak{B}_{x_0}$, since $2\mathfrak{m}(\widehat{x})d(x_0,\mathscr{S})^{-a} \leq d(x_0,\mathscr{S})^a < \mathfrak{d}(x_0)$. By (A1), the composition $(\Psi_{\widehat{x}}^{-1} \circ f_{\widehat{x}}^{-1} \circ \Psi_{\widehat{f(x)}})(v)$ is well-defined. Now we prove parts (1)–(4).

(1) It is clear that, whenever defined, the compositions $F_{\widehat{x}}^{-1} \circ F_{\widehat{x}}$ and $F_{\widehat{x}} \circ F_{\widehat{x}}^{-1}$ are the identity. We start showing that $F_{\widehat{x}}^{-1} \circ F_{\widehat{x}}$ is well-defined in $B[20Q(\widehat{x})]$. By the beginning of the proof, it is enough to show that $F_{\widehat{x}}(B[20Q(\widehat{x})]) \subset B[\mathfrak{m}(\widehat{x})]$. Indeed, assumption (A6) and Lemma 3.2(1) imply that

$$F_{\widehat{x}}(B[20Q(\widehat{x})]) \subset B[80d(x_0,\mathscr{S})^{-a}\|C(\widehat{f}(\widehat{x}))^{-1}\|Q(\widehat{x})]$$

and this latter radius is $80d(x_0,\mathscr{S})^{-a}\|C(\widehat{f}(\widehat{x}))^{-1}\|Q(\widehat{x}) < \frac{1}{2}\rho(\widehat{x})^{2a} \leq \mathfrak{m}(\widehat{x})$ for $\varepsilon > 0$ sufficiently small, since by (3.3) it holds $\rho(\widehat{x})^{-3a}Q(\widehat{x})^{1/2} < \rho(\widehat{x})^{-3a}Q(\widehat{x})^{\beta/32} < \varepsilon^{3/16}$ and $\|C(\widehat{f}(\widehat{x}))^{-1}\|Q(\widehat{x})^{1/2} < \|C(\widehat{f}(\widehat{x}))^{-1}\|Q(\widehat{x})^{\beta/12} < \varepsilon^{1/4}$.

Similarly, $F_{\widehat{x}}^{-1}(B[20Q(\widehat{f}(\widehat{x}))]) \subset B[80d(x_0,\mathscr{S})^{-a}\|C(\widehat{x})^{-1}\|Q(\widehat{f}(\widehat{x}))]$. By Lemma 3.1(4), we have

$$80d(x_0,\mathscr{S})^{-a}\|C(\widehat{x})^{-1}\|Q(\widehat{f}(\widehat{x})) \leq 80(1+e^{2\chi})^{1/2}d(x_0,\mathscr{S})^{-3a}\|C(\widehat{f}(\widehat{x}))^{-1}\|Q(\widehat{f}(\widehat{x}))$$

which is smaller than $\frac{1}{2}\rho(\widehat{f}(\widehat{x}))^{2a}$ since by (3.3) we have $\rho(\widehat{f}(\widehat{x}))^{-5a}Q(\widehat{f}(\widehat{x}))^{1/2} < \varepsilon^{5/16}$ and $\|C(\widehat{f}(\widehat{x}))^{-1}\|Q(\widehat{f}(\widehat{x}))^{1/2} < \varepsilon^{1/8}$.

(2) This follows from the identities $d(\Psi_{\widehat{x}})_0 = C(\widehat{x})$, $df_{x_0} = \widehat{df}_{\widehat{x}}$, $d(f_{\widehat{x}}^{-1})_{x_1} = (df_{x_0})^{-1} = (\widehat{df}_{\widehat{x}})^{-1}$ and from Lemma 3.1(2).

(3) Define $H^+ : B[20Q(\widehat{x})] \to \mathbb{R}^m$ and $H^- : B[20Q(\widehat{f}(\widehat{x}))] \to \mathbb{R}^m$ by the equalities $F_{\widehat{x}} = D(\widehat{x}) + H^+$ and $F_{\widehat{x}}^{-1} = D(\widehat{x})^{-1} + H^-$. Property (a) follows from part (2) above, so it remains to prove (b). We show the following fact.

CLAIM: $\|d(F_{\widehat{x}})_{w_1} - d(F_{\widehat{x}})_{w_2}\| \leq \frac{\varepsilon}{3}\|w_1 - w_2\|^{\beta/2}$ for every $w_1, w_2 \in B[20Q(\widehat{x})]$, and

$\|d(F_{\widehat{x}}^{-1})_{w_1} - d(F_{\widehat{x}}^{-1})_{w_2}\| \leq \frac{\varepsilon}{3}\|w_1 - w_2\|^{\beta/2}$ for every $w_1, w_2 \in B[20Q(\widehat{f}(\widehat{x}))]$.

Before proving the claim, let us show how to conclude (b). Applying the claim with $w_2 = 0$, we get $\|d(H^{\pm})_w\| \leq \frac{\varepsilon}{3}\|w\|^{\beta/2} < \frac{\varepsilon}{3}$ and so $\|dH^{\pm}\|_{C^0} < \frac{\varepsilon}{3}$. By the mean value inequality, $\|H^{\pm}(w)\| \leq \frac{\varepsilon}{3}\|w\| < \frac{\varepsilon}{3}$, thus $\|H^{\pm}\|_{C^0} < \frac{\varepsilon}{3}$. The claim also implies that $\mathrm{Hol}_{\beta/2}(dH^{\pm}) \leq \frac{\varepsilon}{3}$. These estimates imply that $\|H^{\pm}\|_{C^{1+\frac{\beta}{2}}} < \varepsilon$.

PROOF OF THE CLAIM. We begin proving it for $F_{\widehat{x}}$. Remind that $\widehat{x} = (x_n)_{n\in\mathbb{Z}}$. For $i = 1,2$, write $u_i = C(\widehat{x})w_i$ and define

$$A_i = d(\widetilde{\exp_{f(x_0)}^{-1}})_{(f\circ\exp_{x_0})(u_i)},\ B_i = d\widetilde{f_{\exp_{x_0}(u_i)}},\ C_i = d(\widetilde{\exp_{x_0}})_{u_i}.$$

Note that $\|u_1 - u_2\| \leq \|w_1 - w_2\|$. We first estimate $\|A_1B_1C_1 - A_2B_2C_2\|$.

- By (A2), $\|A_i\| \leq 2$. By (A2), (A3), (A6):

$$\begin{aligned}\|A_1 - A_2\| &\leq d(f(x_0), \mathscr{S})^{-a} d((f \circ \exp_{x_0})(u_1), (f \circ \exp_{x_0})(u_2)) \\ &\leq 2d(x_0, \mathscr{S})^{-a} d(f(x_0), \mathscr{S})^{-a} \|u_1 - u_2\| \\ &\leq 2\rho(\widehat{x})^{-2a} \|w_1 - w_2\|.\end{aligned}$$

- By (A6), $\|B_i\| \leq \rho(\widehat{x})^{-a}$. By (A2) and (A7):

$$\|B_1 - B_2\| \leq \mathfrak{K} d(\exp_{x_0}(u_1), \exp_{x_0}(u_2))^\beta \leq 2\mathfrak{K}\|u_1 - u_2\|^\beta \leq 2\mathfrak{K}\|w_1 - w_2\|^\beta.$$

- By (A2), $\|C_i\| \leq 2$. By (A3):

$$\|C_1 - C_2\| \leq d(x_0, \mathscr{S})^{-a} \|u_1 - u_2\| \leq \rho(\widehat{x})^{-a} \|w_1 - w_2\|.$$

Applying some triangle inequalities, we get that

$$\|A_1 B_1 C_1 - A_2 B_2 C_2\| \leq 24\mathfrak{K}\rho(\widehat{x})^{-3a} \|w_1 - w_2\|^\beta.$$

Now we estimate $\|d(F_{\widehat{x}})_{w_1} - d(F_{\widehat{x}})_{w_2}\|$:

$$\begin{aligned}\|d(F_{\widehat{x}})_{w_1} - d(F_{\widehat{x}})_{w_2}\| &\leq \|C(\widehat{f}(\widehat{x}))^{-1}\| \cdot \|A_1 B_1 C_1 - A_2 B_2 C_2\| \cdot \|C(\widehat{x})\| \\ &\leq 24\mathfrak{K}\rho(\widehat{x})^{-3a} \|C(\widehat{f}(\widehat{x}))^{-1}\| \cdot \|w_1 - w_2\|^\beta.\end{aligned}$$

Since $\|w_1 - w_2\| < 40Q(\widehat{x})$, using Lemma 3.1(4) and the definition of $Q(\widehat{x})$ we obtain that for $\varepsilon > 0$ small enough it holds

$$\begin{aligned}&24\mathfrak{K}\rho(\widehat{x})^{-3a} \|C(\widehat{f}(\widehat{x}))^{-1}\| \|w_1 - w_2\|^{\beta/2} \\ &\leq 200\mathfrak{K}(1 + e^{2\chi})^{1/2} \rho(\widehat{x})^{-5a} \|C(\widehat{x})^{-1}\| Q(\widehat{x})^{\beta/2} \\ &= 200\mathfrak{K}(1 + e^{2\chi})^{1/2} \left[\rho(\widehat{x})^{-5a} Q(\widehat{x})^{5\beta/96}\right] \cdot \left[\|C(\widehat{x})^{-1}\| Q(\widehat{x})^{\beta/48}\right] Q(\widehat{x})^{41\beta/96} \\ &\leq 200\mathfrak{K}(1 + e^{2\chi})^{1/2} \varepsilon^{5/16} \varepsilon^{1/8} \varepsilon^{41/16} = 200\mathfrak{K}(1 + e^{2\chi})^{1/2} \varepsilon^3 < \tfrac{\varepsilon}{3},\end{aligned}$$

which completes the proof of the claim for $F_{\widehat{x}}$. Now we prove it for $F_{\widehat{x}}^{-1}$. For $i = 1, 2$, write $u_i = C(\widehat{f}(\widehat{x}))w_i$ and define

$$A_i = \widetilde{d(\exp_{x_0}^{-1})}_{(f_{\widehat{x}}^{-1} \circ \exp_{f(x_0)})(u_i)}, \quad B_i = \widetilde{d(f_{\widehat{x}}^{-1})}_{\exp_{f(x_0)}(u_i)}, \quad C_i = \widetilde{d(\exp_{f(x_0)})}_{u_i}.$$

Again, we have $\|u_1 - u_2\| \leq \|w_1 - w_2\|$ and the following similar estimates:

- By (A2), $\|A_i\| \leq 2$. By (A2), (A3), (A6):

$$\begin{aligned}\|A_1 - A_2\| &\leq d(x_0, \mathscr{S})^{-a} d((f_{\widehat{x}}^{-1} \circ \exp_{f(x_0)})(u_1), (f_{\widehat{x}}^{-1} \circ \exp_{f(x_0)})(u_2)) \\ &\leq 2d(x_0, \mathscr{S})^{-2a} \|u_1 - u_2\| \\ &\leq 2\rho(\widehat{f}(\widehat{x}))^{-2a} \|w_1 - w_2\|.\end{aligned}$$

- By (A6), $\|B_i\| \leq \rho(\widehat{f}(\widehat{x}))^{-a}$. By (A2) and (A7):

$$\|B_1 - B_2\| \leq \mathfrak{K} d(\exp_{f(x_0)}(u_1), \exp_{f(x_0)}(u_2))^\beta \leq 2\mathfrak{K}\|u_1 - u_2\|^\beta \leq 2\mathfrak{K}\|w_1 - w_2\|^\beta.$$

- By (A2), $\|C_i\| \leq 2$. By (A3):

$$\|C_1 - C_2\| \leq d(f(x_0), \mathscr{S})^{-a} \|u_1 - u_2\| \leq \rho(\widehat{f}(\widehat{x}))^{-a} \|w_1 - w_2\|.$$

As before, the above estimates imply that

$$\|A_1B_1C_1 - A_2B_2C_2\| \leq 24\mathfrak{K}\rho(\widehat{f}(\widehat{x}))^{-3a}\|w_1 - w_2\|^\beta$$

and then a calculation identical to the one made for $F_{\widehat{x}}$ works:

$$\begin{aligned}
&24\mathfrak{K}\rho(\widehat{f}(\widehat{x}))^{-3a}\|C(\widehat{x})^{-1}\|\|w_1 - w_2\|^{\beta/2}\\
&\leq 200\mathfrak{K}(1+e^{2\chi})^{1/2}\rho(\widehat{f}(\widehat{x}))^{-5a}\|C(\widehat{f}(\widehat{x}))^{-1}\|Q(\widehat{f}(\widehat{x}))^{\beta/2}\\
&\leq 200\mathfrak{K}(1+e^{2\chi})^{1/2}\varepsilon^3 < \tfrac{\varepsilon}{3}.
\end{aligned}$$

Hence $\|d(F_{\widehat{x}}^{-1})_{w_1} - d(F_{\widehat{x}}^{-1})_{w_2}\| < \frac{\varepsilon}{3}\|w_1 - w_2\|^{\beta/2}$, thus proving the claim. □

(4) By Lemma 3.1(3) and part (2) above,

$$\|d(F_{\widehat{x}})_0\| = \|D_u(\widehat{x})\| \leq d(\vartheta[\widehat{x}], \mathscr{S})^{-a}(1+e^{2\chi})^{1/2}.$$

By part (3) above, if $w \in B[20Q(\widehat{x})]$ then

$$\|d(F_{\widehat{x}})_w\| \leq \varepsilon\|w\|^{\beta/2} + d(\vartheta[\widehat{x}], \mathscr{S})^{-a}(1+e^{2\chi})^{1/2} < 2d(\vartheta[\widehat{x}], \mathscr{S})^{-a}(1+e^{2\chi})^{1/2},$$

since $\varepsilon\|w\|^{\beta/2} < 1 < d(\vartheta[\widehat{x}], \mathscr{S})^{-a}(1+e^{2\chi})^{1/2}$ for small $\varepsilon > 0$. Similarly, using that $\|d(F_{\widehat{x}}^{-1})_0\| = \|D_s(\widehat{x})^{-1}\| \leq d(\vartheta[\widehat{x}], \mathscr{S})^{-a}(1+e^{2\chi})^{1/2}$ we conclude that $\|d(F_{\widehat{x}}^{-1})_w\| < 2d(\vartheta[\widehat{x}], \mathscr{S})^{-a}(1+e^{2\chi})^{1/2}$ for all $w \in B[20Q(\widehat{f}(\widehat{x}))]$. □

An analogous result holds for the Lyapunov inner product $|\!|\!|\cdot|\!|\!|_{\chi'}$: defining $F_{\chi',\widehat{x}}$ and $F_{\chi',\widehat{x}}^{-1}$ as compositions with the Pesin charts $\Psi_{\chi',\cdot}$, then Theorem 3.3 holds with $D(\widehat{x})$ and χ replaced by $D_{\chi'}(\widehat{x})$ and χ', but still in the same domain $B[20Q(\widehat{x})]$ (indeed, it holds in the larger domain $B[20Q_{\chi'}(\widehat{x})]$).

3.4. Temperedness and the parameters $q(\widehat{x}), q^s(\widehat{x}), q^u(\widehat{x})$

Now we will focus on the points for which Q does not converge exponentially fast to zero too fast along the trajectory of $\widehat{x}$. Intuitively, this condition will guarantee that hyperbolicity will prevail over the degeneracies of nonuniform hyperbolicity, of the derivative, and of the proximity to $\mathscr{S}$. Recall that $\chi > 0$ is fixed a priori and $\varepsilon > 0$ is sufficiently small. Define $\delta_\varepsilon := e^{-\varepsilon n} \in I_\varepsilon$ where n is the unique positive integer s.t. $e^{-\varepsilon n} < \varepsilon \leq e^{-\varepsilon(n-1)}$. In particular, we have $\delta_\varepsilon < \varepsilon$.

PARAMETER $q(\widehat{x})$: For $\widehat{x} \in \mathrm{NUH}$, let $q(\widehat{x}) := \delta_\varepsilon \min\{e^{\varepsilon|n|}Q(\widehat{f}^n(\widehat{x})) : n \in \mathbb{Z}\}$.

THE SET NUH^*: It is the set of $\widehat{x} \in \mathrm{NUH}$ such that $q(\widehat{x}) > 0$.

Observe that if $\widehat{x} \in \mathrm{NUH}$ satisfies

$$\lim_{n\to\pm\infty} \tfrac{1}{|n|}\log\|C(\widehat{f}^n(\widehat{x}))^{-1}\| = \lim_{n\to\pm\infty} \tfrac{1}{|n|}\log d(\vartheta_n[\widehat{x}], \mathscr{S}) = 0,$$

then $\lim_{n\to\pm\infty} \frac{1}{|n|}\log Q(\widehat{f}^n(\widehat{x})) = 0$ and so $\widehat{x} \in \mathrm{NUH}^*$. The next lemma states the basic properties of q.

LEMMA 3.4. *For all $\widehat{x} \in \mathrm{NUH}^*$, $0 < q(\widehat{x}) < \varepsilon Q(\widehat{x})$ and $\frac{q(\widehat{f}(\widehat{x}))}{q(\widehat{x})} = e^{\pm\varepsilon}$.*

This is [**LM18**, Lemma 4.1], and the proof goes through without change. Now we introduce two more parameters, that will be regarded as sizes of the stable and unstable manifolds at $\widehat{x}$.

PARAMETERS $q^s(\widehat{x}), q^u(\widehat{x})$: For $\widehat{x} \in \mathrm{NUH}$, define

$$q^s(\widehat{x}) := \delta_\varepsilon \min\{e^{\varepsilon|n|} Q(\widehat{f}^n(\widehat{x})) : n \geq 0\},$$
$$q^u(\widehat{x}) := \delta_\varepsilon \min\{e^{\varepsilon|n|} Q(\widehat{f}^n(\widehat{x})) : n \leq 0\}.$$

Note that $q^s(\widehat{x}), q^u(\widehat{x})$ are just the separation, in the definition of $q(\widehat{x})$, of the past from the future. We list the main properties of $q^s(\widehat{x})$ and $q^u(\widehat{x})$.

LEMMA 3.5. *For all* $\widehat{x} \in \mathrm{NUH}^*$, *the following holds.*

(1) GOOD DEFINITION: $0 < q^s(\widehat{x}), q^u(\widehat{x}) < \varepsilon Q(\widehat{x})$ *and* $q^s(\widehat{x}) \wedge q^u(\widehat{x}) = q(\widehat{x})$.
(2) GREEDY ALGORITHM: *For all* $n \in \mathbb{Z}$ *it holds*

$$q^s(\widehat{f}^n(\widehat{x})) = \min\{e^\varepsilon q^s(\widehat{f}^{n+1}(\widehat{x})), \delta_\varepsilon Q(\widehat{f}^n(\widehat{x}))\},$$
$$q^u(\widehat{f}^n(\widehat{x})) = \min\{e^\varepsilon q^u(\widehat{f}^{n-1}(\widehat{x})), \delta_\varepsilon Q(\widehat{f}^n(\widehat{x}))\}.$$

This is [**LM18**, Lemma 4.2], and the same proof works. Now we introduce the set that we will effectively code.

THE SET $\mathrm{NUH}^\#$: It is the set of $\widehat{x} \in \mathrm{NUH}^*$ s.t.

$$\limsup_{n\to+\infty} q(\widehat{f}^n(\widehat{x})) > 0 \text{ and } \limsup_{n\to-\infty} q(\widehat{f}^n(\widehat{x})) > 0.$$

For a fixed $t > 0$, the set $\{\widehat{x} \in \mathrm{NUH}^* : q(\widehat{x}) > t\}$ can be regarded, in the usual literature, as a Pesin block. In this sense, $\widehat{x} \in \mathrm{NUH}^\#$ iff its trajectory returns to some Pesin block infinitely often in the past and in the future. Therefore, $\mathrm{NUH}^\#$ is the dynamical counterpart of the recurrent set $\Sigma^\#$ introduced in Section 2.1. The next lemma shows that $\mathrm{NUH}^\#$ is large for all f-adapted χ-hyperbolic measures.

LEMMA 3.6. *If* μ *is an* f*-adapted measure and* $\widehat{\mu}$ *is carried by* NUH, *then* $\widehat{\mu}$ *is carried by* $\mathrm{NUH}^\#$. *In particular,* $\mathrm{NUH}^\#$ *carries all* f*-adapted* χ*-hyperbolic measures.*

PROOF. Fix an f-adapted measure μ s.t. $\widehat{\mu}[\mathrm{NUH}] = 1$. By the Poincaré recurrence theorem, every measure supported in NUH^* is supported in $\mathrm{NUH}^\#$, so it is enough to show that

$$\lim_{n\to\pm\infty} \tfrac{1}{|n|} \log \|C(\widehat{f}^n(\widehat{x}))^{-1}\| = \lim_{n\to\pm\infty} \tfrac{1}{|n|} \log \rho(\widehat{f}^n(\widehat{x})) = 0$$

for $\widehat{\mu}$-a.e. $\widehat{x} \in \mathrm{NUH}$. Let us prove these. Since μ is f-adapted, $\log\rho \in L^1(\widehat{\mu})$ and so by the Birkhoff ergodic theorem $\lim\limits_{n\to\pm\infty} \frac{1}{|n|} \log \rho(\widehat{f}^n(\widehat{x})) = 0$ for $\widehat{\mu}$-a.e. $\widehat{x} \in \mathrm{NUH}$. For the other equality, define $\varphi : \mathrm{NUH} \to \mathbb{R}$ by

$$\varphi(\widehat{x}) := \log\left[\tfrac{\|C(\widehat{f}(\widehat{x}))^{-1}\|}{\|C(\widehat{x})^{-1}\|}\right] = \log\|C(\widehat{f}(\widehat{x}))^{-1}\| - \log\|C(\widehat{x})^{-1}\|.$$

Using Lemma 3.1(4) and again that $\log\rho \in L^1(\widehat{\mu})$, we get that φ is in $L^1(\widehat{\mu})$. Letting $\varphi_n(\widehat{x}) = \log\|C(\widehat{f}^n(\widehat{x}))^{-1}\| - \log\|C(\widehat{x})^{-1}\|$ denote the nth Birkhoff sum of φ, by the Birkhoff ergodic theorem $\lim\limits_{n\to+\infty} \frac{\varphi_n(\widehat{x})}{n}$ exists $\widehat{\mu}$-a.e. The Poincaré recurrence theorem implies that

$$\liminf_{n\to+\infty} |\varphi_n(\widehat{x})| = \liminf_{n\to+\infty} \left|\log\|C(\widehat{f}^n(\widehat{x}))^{-1}\| - \log\|C(\widehat{x})^{-1}\|\right| < \infty$$

for $\widehat{\mu}$-a.e $\widehat{x} \in \text{NUH}$, hence $\lim\limits_{n\to+\infty} \frac{\varphi_n(\widehat{x})}{n} = 0$ for $\widehat{\mu}$-a.e. $\widehat{x} \in \text{NUH}$. Proceeding similarly for $n \to -\infty$, we get that $\lim_{n\to\pm\infty} \frac{1}{|n|} \log \|C(\widehat{f}^n(\widehat{x}))^{-1}\| = 0$ for $\widehat{\mu}$-a.e. $\widehat{x} \in \text{NUH}$. □

CHAPTER 4

Overlap, double charts, and graph transforms

We will use the Pesin charts to apply the graph transform method. For that, we allow the size of the domain of definition to vary. Recall that $B[r]$ is the ball of center 0 and radius r in $\mathbb{R}^m$.

PESIN CHART $\Psi^{\eta}_{\widehat{x}}$: It is the restriction of $\Psi_{\widehat{x}}$ to $B[\eta]$, where $0 < \eta \leq Q(\widehat{x})$.

4.1. The overlap condition

We recall a notion, introduced in [**Sar13**], that allows to change coordinates from $\Psi_{\widehat{x}_1}$ to $\Psi_{\widehat{x}_2}$.

ε-OVERLAP: Two Pesin charts $\Psi^{\eta_1}_{\widehat{x}_1}, \Psi^{\eta_2}_{\widehat{x}_2}$ are said to *ε-overlap* if $d_{s/u}(\widehat{x}_1) = d_{s/u}(\widehat{x}_2)$, $\frac{\eta_1}{\eta_2} = e^{\pm\varepsilon}$ and if $\exists x \in M$ s.t. $\vartheta[\widehat{x}_1], \vartheta[\widehat{x}_2] \in \mathfrak{B}_x$ and

$$d(\vartheta[\widehat{x}_1], \vartheta[\widehat{x}_2]) + \left\|\widetilde{C(\widehat{x}_1)} - \widetilde{C(\widehat{x}_2)}\right\| < (\eta_1\eta_2)^4.$$

When this happens, we write $\Psi^{\eta_1}_{\widehat{x}_1} \overset{\varepsilon}{\approx} \Psi^{\eta_2}_{\widehat{x}_2}$.

We claim that if $\varepsilon > 0$ is small enough, then $\Psi^{\eta_1}_{\widehat{x}_1} \overset{\varepsilon}{\approx} \Psi^{\eta_2}_{\widehat{x}_2}$ implies that

$$\Psi_{\widehat{x}_i}(B[20Q(\widehat{x}_i)]) \subset \mathfrak{B}_{\vartheta[\widehat{x}_1]} \cap \mathfrak{B}_{\vartheta[\widehat{x}_2]} \text{ for } i = 1,2 \tag{4.1}$$

and hence we can apply (A1)–(A4) without mentioning x. We prove this for $i = 1$. Firstly, since $d(\vartheta[\widehat{x}_1], \vartheta[\widehat{x}_2]) < \varepsilon^2 d(\vartheta[\widehat{x}_2], \mathscr{S})$, we have $d(\vartheta[\widehat{x}_1], \mathscr{S}) = d(\vartheta[\widehat{x}_2], \mathscr{S}) \pm d(\vartheta[\widehat{x}_1], \vartheta[\widehat{x}_2]) = (1 \pm \varepsilon^2) d(\vartheta[\widehat{x}_2], \mathscr{S}) = e^{\pm\varepsilon} d(\vartheta[\widehat{x}_2], \mathscr{S})$. By Lemma 3.2(1), $\Psi_{\widehat{x}_1}(B[20Q(\widehat{x}_1)]) \subset B(\vartheta[\widehat{x}_1], 40Q(\widehat{x}_1))$ which is contained in $\mathfrak{B}_{\vartheta[\widehat{x}_1]}$ since $40Q(\widehat{x}_1) \ll 40\varepsilon^{6/\beta}\rho(\widehat{x}_1)^a < 2\mathfrak{d}(\widehat{x}_1)$. For the other inclusion, note that

$$\Psi_{\widehat{x}_1}(B[20Q(\widehat{x}_1)]) \subset B(\vartheta[\widehat{x}_1], 40Q(\widehat{x}_1)) \subset B(\vartheta[\widehat{x}_2], 40Q(\widehat{x}_1) + d(\vartheta[\widehat{x}_1], \vartheta[\widehat{x}_2])).$$

Since $40Q(\widehat{x}_1) + d(\vartheta[\widehat{x}_1], \vartheta[\widehat{x}_2]) \leq 40\varepsilon^{6/\beta} e^{a\varepsilon} d(\vartheta[\widehat{x}_2], \mathscr{S})^a + d(\vartheta[\widehat{x}_2], \mathscr{S})^a < 2\mathfrak{d}(\widehat{x}_2)$ for $\varepsilon > 0$ small enough, it follows that $\Psi_{\widehat{x}_1}(B[20Q(\widehat{x}_1)]) \subset \mathfrak{B}_{\vartheta[\widehat{x}_2]}$.

The next result quantifies the closeness of Pesin charts when an ε-overlap occurs.

PROPOSITION 4.1. *The following holds for $\varepsilon > 0$ small enough. If $\Psi^{\eta_1}_{\widehat{x}_1} \overset{\varepsilon}{\approx} \Psi^{\eta_2}_{\widehat{x}_2}$ and $C_i = \widetilde{C(\widehat{x}_i)}$ for $i = 1,2$ then:*

(1) CONTROL OF ρ: $\frac{d(\vartheta_i[\widehat{x}_2], \mathscr{S})}{d(\vartheta_i[\widehat{x}_1], \mathscr{S})} = e^{\pm\varepsilon}$ *for $i = -1, 0, 1$. In particular,* $\frac{\rho(\widehat{x}_1)}{\rho(\widehat{x}_2)} = e^{\pm\varepsilon}$.

(2) CONTROL OF C^{-1}: $\left\|C_1^{-1} - C_2^{-1}\right\| < (\eta_1\eta_2)^3$ *and* $\frac{\left\|C_1^{-1}\right\|}{\left\|C_2^{-1}\right\|} = e^{\pm(\eta_1\eta_2)^3}$.

(3) CONTROL OF Q: $\frac{Q(\widehat{x}_1)}{Q(\widehat{x}_2)} = \exp[\pm\frac{97a\varepsilon}{\beta}]$.

(4) OVERLAP: $\Psi_{\widehat{x}_i}(B[e^{-2\varepsilon}\eta_i]) \subset \Psi_{\widehat{x}_j}(B[\eta_j])$ *for* $i, j = 1, 2$.
(5) CHANGE OF COORDINATES: *For* $i, j = 1, 2$, *the map* $\Psi_{\widehat{x}_i}^{-1} \circ \Psi_{\widehat{x}_j}$ *is well-defined in* $B[d(\vartheta[\widehat{x}_j], \mathscr{S})^a]$, *and* $\left\|\Psi_{\widehat{x}_i}^{-1} \circ \Psi_{\widehat{x}_j} - \mathrm{Id}\right\|_{C^{1+\frac{\beta}{2}}} < \varepsilon(\eta_1\eta_2)^2$ *where the norm is taken in* $B[d(\vartheta[\widehat{x}_j], \mathscr{S})^{2a}]$.

PROOF. Recall from (3.3) that $\|C_i^{-1}\| \leq \varepsilon^{1/8} Q(\widehat{x}_i)^{-\beta/48} \leq \varepsilon^{1/8}\eta_i^{-\beta/48}$.

(1) The case $i = 0$ was treated above. Consider $i = 1$. Since $d(\vartheta[\widehat{x}_1], \vartheta[\widehat{x}_2]) < (\eta_1\eta_2)^4 < \varepsilon^2\rho(\widehat{x}_2)^{2a}$, we have that $\vartheta[\widehat{x}_1] \in D_{\widehat{x}_2}$. By (A6), $d(\vartheta_1[\widehat{x}_1], \vartheta_1[\widehat{x}_2]) \leq d(\vartheta[\widehat{x}_2], \mathscr{S})^{-a} d(\vartheta[\widehat{x}_1], \vartheta[\widehat{x}_2]) < \varepsilon^2\rho(\widehat{x}_2)^a < \varepsilon^2 d(\vartheta_1[\widehat{x}_2], \mathscr{S})$ and so $d(\vartheta_1[\widehat{x}_1], \mathscr{S}) = d(\vartheta_1[\widehat{x}_2], \mathscr{S}) \pm d(\vartheta_1[\widehat{x}_1], \vartheta_1[\widehat{x}_2]) = (1 \pm \varepsilon^2) d(\vartheta_1[\widehat{x}_2], \mathscr{S}) = e^{\pm\varepsilon} d(\vartheta_1[\widehat{x}_2], \mathscr{S})$. The same proof applies to $i = -1$.

(2) We have $C_1^{-1} - C_2^{-1} = C_1^{-1}(C_2 - C_1)C_2^{-1}$, hence

$$\|C_1^{-1} - C_2^{-1}\| \leq \|C_1^{-1}\| \cdot \|C_2^{-1}\| \cdot \|C_1 - C_2\| \leq \varepsilon^{1/4}(\eta_1\eta_2)^{4-\beta/48} < \tfrac{1}{2}(\eta_1\eta_2)^3.$$

Additionally,

$$\left|\frac{\|C_1^{-1}\|}{\|C_2^{-1}\|} - 1\right| \leq \|C_1^{-1} - C_2^{-1}\| \leq \tfrac{1}{2}(\eta_1\eta_2)^3$$

and so $\frac{\|C_1^{-1}\|}{\|C_2^{-1}\|} = e^{\pm(\eta_1\eta_2)^3}$.

(3) By parts (1) and (2), $\frac{\rho(\widehat{x}_1)^{96a/\beta}}{\rho(\widehat{x}_2)^{96a/\beta}} = \exp\left[\pm\frac{96a\varepsilon}{\beta}\right]$, and

$$\frac{\|C(\widehat{x}_1)^{-1}\|^{-48/\beta}}{\|C(\widehat{x}_2)^{-1}\|^{-48/\beta}} = \exp\left[\pm\frac{48\varepsilon}{\beta}\right].$$

Since $\frac{48\varepsilon}{\beta} < \frac{96a\varepsilon}{\beta}$, we get $\frac{\widetilde{Q}(\widehat{x}_1)}{\widetilde{Q}(\widehat{x}_2)} = \exp\left[\pm\frac{96a\varepsilon}{\beta}\right]$. By the definition of Q, we obtain

$$\frac{Q(\widehat{x}_1)}{Q(\widehat{x}_2)} = \exp\left[\pm\left(\frac{96a\varepsilon}{\beta} + \frac{\varepsilon}{3}\right)\right] = \exp\left[\pm\frac{97a\varepsilon}{\beta}\right].$$

(4) The proof of this and of the next part follows [**LM18**, Proposition 3.4]. We prove that $\Psi_{\widehat{x}_1}(B[e^{-2\varepsilon}\eta_1]) \subset \Psi_{\widehat{x}_2}(B[\eta_2])$. If $v \in B[e^{-2\varepsilon}\eta_1]$ then $\|C(\widehat{x}_1)v\| \leq e^{-2\varepsilon}\eta_1 < 2\mathfrak{d}(x)$, hence by (A1):

$$d_{\mathrm{Sas}}(C(\widehat{x}_1)v, C(\widehat{x}_2)v) \leq 2\left[d(\vartheta[\widehat{x}_1], \vartheta[\widehat{x}_2]) + \|C_1 v - C_2 v\|\right] \leq 2(\eta_1\eta_2)^4.$$

By (A2), $d(\Psi_{\widehat{x}_1}(v), \Psi_{\widehat{x}_2}(v)) \leq 4(\eta_1\eta_2)^4$ and so $\Psi_{\widehat{x}_1}(v) \in B(\Psi_{\widehat{x}_2}(v), 4(\eta_1\eta_2)^4)$. By Lemma 3.2(1), $B(\Psi_{\widehat{x}_2}(v), 4(\eta_1\eta_2)^4) \subset \Psi_{\widehat{x}_2}(B)$ where $B \subset \mathbb{R}^m$ is the ball with center v and radius $8\|C_2^{-1}\|(\eta_1\eta_2)^4$, hence it is enough that $B \subset B[\eta_2]$. If $w \in B$ then $\|w\| \leq \|v\| + 8\|C_2^{-1}\|(\eta_1\eta_2)^4 \leq (e^{-\varepsilon} + 8\varepsilon^{6/\beta})\eta_2 < \eta_2$ for $\varepsilon > 0$ small enough.

(5) The proof that $\Psi_{\widehat{x}_2}^{-1} \circ \Psi_{\widehat{x}_1}$ is well-defined in $B[d(\vartheta[\widehat{x}_1], \mathscr{S})^a]$ is similar to the proof of (4), the only difference being in the last estimate: if $\varepsilon > 0$ is small enough then for $w \in B$ it holds

$$\begin{aligned}
\|w\| &\leq \|v\| + 8\|C_2^{-1}\|(\eta_1\eta_2)^4 \\
&\leq d(\vartheta[\widehat{x}_1], \mathscr{S})^a + 8(\eta_1\eta_2)^3 \\
&\leq [e^{a\varepsilon} + 8\varepsilon^{6/\beta}] d(\vartheta[\widehat{x}_2], \mathscr{S})^a < 2\mathfrak{d}(\widehat{x}_2).
\end{aligned}$$

Now:

$$\begin{aligned}\Psi_{\widehat{x}_2}^{-1} \circ \Psi_{\widehat{x}_1} - \mathrm{Id} &= C(\widehat{x}_2)^{-1} \circ \exp_{\vartheta[\widehat{x}_2]}^{-1} \circ \exp_{\vartheta[\widehat{x}_1]} \circ C(\widehat{x}_1) - \mathrm{Id} \\ &= [C_2^{-1} \circ P_{\vartheta[\widehat{x}_2],x}] \circ [\exp_{\vartheta[\widehat{x}_2]}^{-1} \circ \exp_{\vartheta[\widehat{x}_1]} - P_{\vartheta[\widehat{x}_1],\vartheta[\widehat{x}_2]}] \circ [P_{x,\vartheta[\widehat{x}_1]} \circ C_1] \\ &\quad + C_2^{-1}(C_1 - C_2) \\ &= [C_2^{-1} \circ P_{\vartheta[\widehat{x}_2],x}] \circ [\exp_{\vartheta[\widehat{x}_2]}^{-1} - P_{\vartheta[\widehat{x}_1],\vartheta[\widehat{x}_2]} \circ \exp_{\vartheta[\widehat{x}_1]}^{-1}] \circ \Psi_{\widehat{x}_1} + C_2^{-1}(C_1 - C_2).\end{aligned}$$

We calculate the $C^{1+\frac{\beta}{2}}$ norm of $[\exp_{\vartheta[\widehat{x}_2]}^{-1} - P_{\vartheta[\widehat{x}_1],\vartheta[\widehat{x}_2]} \circ \exp_{\vartheta[\widehat{x}_1]}^{-1}] \circ \Psi_{\widehat{x}_1}$ in the ball $B[d(\vartheta[\widehat{x}_1],\mathscr{S})^{2a}]$. By Lemma 3.2, $\|d\Psi_{\widehat{x}_1}\|_0 \leq 2$ and

$$\mathrm{Hol}_{\frac{\beta}{2}}(d\Psi_{\widehat{x}_1}) \leq d(\vartheta[\widehat{x}_1],\mathscr{S})^{-a} 2d(\vartheta[\widehat{x}_1],\mathscr{S})^{2a(1-\beta/2)} = 2d(\vartheta[\widehat{x}_1],\mathscr{S})^{a(1-\beta)} < 2.$$

Call $\Theta := \exp_{\vartheta[\widehat{x}_2]}^{-1} - P_{\vartheta[\widehat{x}_1],\vartheta[\widehat{x}_2]} \circ \exp_{\vartheta[\widehat{x}_1]}^{-1}$. For $\varepsilon > 0$ sufficiently small, inside $\mathfrak{B}_{\widehat{x}_1}$ we have the following bounds:

- By (A2), $\|\Theta(z)\| \leq d_{\mathrm{Sas}}(\exp_{\vartheta[\widehat{x}_2]}^{-1}(z), \exp_{\vartheta[\widehat{x}_1]}^{-1}(z)) \leq 2d(\vartheta[\widehat{x}_1],\vartheta[\widehat{x}_2]) \leq 2\varepsilon^{6/\beta}(\eta_1\eta_2)^3$ thus $\|\Theta \circ \Psi_{\widehat{x}_1}\|_{C^0} < \varepsilon^{2/\beta}(\eta_1\eta_2)^3$.
- By (A3),

$$\begin{aligned}\|d\Theta_z\| &= \|\tau(\vartheta[\widehat{x}_2], z) - \tau(\vartheta[\widehat{x}_1], z)\| \\ &\leq d(\vartheta[\widehat{x}_1],\mathscr{S})^{-a} d(\vartheta[\widehat{x}_1],\vartheta[\widehat{x}_2]) \\ &< \varepsilon^{6/\beta}(\eta_1\eta_2)^3.\end{aligned}$$

  Hence $\|d\Theta\|_{C^0} < \varepsilon^{6/\beta}(\eta_1\eta_2)^3$ and $\|d(\Theta \circ \Psi_{\widehat{x}_1})\|_{C^0} \leq 2\varepsilon^{6/\beta}(\eta_1\eta_2)^3 < \varepsilon^{2/\beta}(\eta_1\eta_2)^3$.
- By (A4),

$$\begin{aligned}\left\|\widetilde{d\Theta_y} - \widetilde{d\Theta_z}\right\| &= \|[\tau(\vartheta[\widehat{x}_2], y) - \tau(\vartheta[\widehat{x}_1], y)] - [\tau(\vartheta[\widehat{x}_2], z) - \tau(\vartheta[\widehat{x}_1], z)]\| \\ &\leq d(\vartheta[\widehat{x}_1],\mathscr{S})^{-a} d([\vartheta[\widehat{x}_1],\vartheta[\widehat{x}_2]) d(y,z)\end{aligned}$$

  hence $\mathrm{Lip}\,(d\Theta) \leq d(\vartheta[\widehat{x}_1],\mathscr{S})^{-a} d(\vartheta[\widehat{x}_1],\vartheta[\widehat{x}_2])$.
- Using that

$$\begin{aligned}\mathrm{Hol}_{\frac{\beta}{2}}(d(\Theta_1 \circ \Theta_2)) \leq \|d\Theta_1\|_{C^0} \mathrm{Hol}_{\frac{\beta}{2}}(d\Theta_2) + \\ \mathrm{Lip}\,(d\Theta_1)\, \|d\Theta_2\|_{C^0}^2 2d(\vartheta[\widehat{x}_1],\mathscr{S})^{2a(1-\beta/2)}\end{aligned}$$

  for Θ_2 defined in $B[d(\vartheta[\widehat{x}_1],\mathscr{S})^{2a}]$, we obtain that

$$\begin{aligned}&\mathrm{Hol}_{\frac{\beta}{2}}[d(\Theta \circ \Psi_{\widehat{x}_1})] \\ &\leq \|d\Theta\|_{C^0} \mathrm{Hol}_{\frac{\beta}{2}}(d\Psi_{\widehat{x}_1}) + \mathrm{Lip}\,(d\Theta)\, \|d\Psi_{\widehat{x}_1}\|_{C^0}^2 2d(\vartheta[\widehat{x}_1],\mathscr{S})^{2a(1-\beta/2)} \\ &< 2\varepsilon^{6/\beta}(\eta_1\eta_2)^3 + d(\vartheta[\widehat{x}_1],\mathscr{S})^{-a} d(\vartheta[\widehat{x}_1],\vartheta[\widehat{x}_2]) 8d(\vartheta[\widehat{x}_1],\mathscr{S})^{2a(1-\beta/2)} \\ &< 2\varepsilon^{6/\beta}(\eta_1\eta_2)^3 + 8\varepsilon^{6/\beta}(\eta_1\eta_2)^3 < \varepsilon^{2/\beta}(\eta_1\eta_2)^3.\end{aligned}$$

This implies that $\|\Theta \circ \Psi_{\widehat{x}_1}\|_{C^{1+\frac{\beta}{2}}} < 3\varepsilon^{2/\beta}(\eta_1\eta_2)^3$, hence

$$\left\|C_2^{-1} \circ P_{\vartheta[\widehat{x}_2],x} \circ \Theta \circ \Psi_{\widehat{x}_1}\right\|_{C^{1+\frac{\beta}{2}}} \leq \|C_2^{-1}\| 3\varepsilon^{2/\beta}(\eta_1\eta_2)^3 \leq 3\varepsilon^{2/\beta}(\eta_1\eta_2)^2.$$

Thus $\left\|\Psi_{\widehat{x}_2}^{-1} \circ \Psi_{\widehat{x}_1} - \mathrm{Id}\right\|_{C^{1+\frac{\beta}{2}}} \leq 3\varepsilon^{2/\beta}(\eta_1\eta_2)^2 + \|C_2^{-1}\|(\eta_1\eta_2)^4 < 3\varepsilon^{2/\beta}(\eta_1\eta_2)^2 + \varepsilon^{6/\beta}(\eta_1\eta_2)^3 < 4\varepsilon^{2/\beta}(\eta_1\eta_2)^2 < \varepsilon(\eta_1\eta_2)^2$. □

4.2. The maps $F_{\widehat{x},\widehat{y}}$ and $F^{-1}_{\widehat{x},\widehat{y}}$

Let $\widehat{x},\widehat{y} \in \mathrm{NUH}$. When $\Psi^{\eta}_{\widehat{f}(\widehat{x})} \overset{\varepsilon}{\approx} \Psi^{\eta'}_{\widehat{y}}$, we are able to change $\Psi_{\widehat{f}(\widehat{x})}$ by $\Psi_{\widehat{y}}$ in $F_{\widehat{x}}$. Similarly, when $\Psi^{\eta}_{\widehat{x}} \overset{\varepsilon}{\approx} \Psi^{\eta'}_{\widehat{f}^{-1}(\widehat{y})}$, we can change $\Psi_{\widehat{x}}$ by $\Psi_{\widehat{f}^{-1}(\widehat{y})}$ in $F^{-1}_{\widehat{x}}$. In this section, we prove that these changes define maps that satisfy a result similar to Theorem 3.3.

THE MAPS $F_{\widehat{x},\widehat{y}}$ AND $F^{-1}_{\widehat{x},\widehat{y}}$: If $\Psi^{\eta}_{\widehat{f}(\widehat{x})} \overset{\varepsilon}{\approx} \Psi^{\eta'}_{\widehat{y}}$, define $F_{\widehat{x},\widehat{y}} := \Psi^{-1}_{\widehat{y}} \circ f \circ \Psi_{\widehat{x}}$. Similarly, if $\Psi^{\eta}_{\widehat{x}} \overset{\varepsilon}{\approx} \Psi^{\eta'}_{\widehat{f}^{-1}(\widehat{y})}$, define $F^{-1}_{\widehat{x},\widehat{y}} := \Psi^{-1}_{\widehat{x}} \circ f^{-1}_{\widehat{x}} \circ \Psi_{\widehat{y}}$.

We will write $F^{\pm}_{\widehat{x},\widehat{y}}$ to represent both $F_{\widehat{x},\widehat{y}}$ and $F^{-1}_{\widehat{x},\widehat{y}}$. In the above form, the definitions are not symmetric, since the natural would be to define $F^{-1}_{\widehat{x},\widehat{y}}$ taking the inverse branch $f^{-1}_{\widehat{f}^{-1}(\widehat{y})}$ instead of $f^{-1}_{\widehat{x}}$. As a matter of fact, the overlap is strong enough to guarantee that, in our domains of definition, $f^{-1}_{\widehat{x}}$ and $f^{-1}_{\widehat{f}^{-1}(\widehat{y})}$ coincide. We state this as a separate lemma, since it will be used a few times in the text.

LEMMA 4.2. *Let $\widehat{x},\widehat{y} \in \mathrm{NUH}$ and assume $\Psi^{\eta}_{\widehat{x}} \overset{\varepsilon}{\approx} \Psi^{\eta'}_{\widehat{f}^{-1}(\widehat{y})}$. If $A = \Psi_{\widehat{y}}(B[20Q(\widehat{y})])$, then $A \subset E_{\widehat{x}}$ and $f^{-1}_{\widehat{f}^{-1}(\widehat{y})}(A) \subset f^{-1}_{\widehat{x}}(E_{\widehat{x}})$.*

PROOF. Write $\widehat{x} = (x_n)_{n\in\mathbb{Z}}$ and $\widehat{y} = (y_n)_{n\in\mathbb{Z}}$. The map $f^{-1}_{\widehat{x}}$ is the inverse branch of f taking x_1 to x_0 and, by (A5), it is well-defined in E_{x_0}. By Proposition 4.1(1), we have $d(y_{-1},\mathscr{S}) = e^{\pm\varepsilon}d(x_0,\mathscr{S})$ and $d(y_0,\mathscr{S}) = e^{\pm\varepsilon}d(x_1,\mathscr{S})$, then

$$\begin{aligned} Q(\widehat{y}) &< \varepsilon^{6/\beta}\rho(\widehat{y})^a \le \varepsilon^{6/\beta}\min\{d(y_{-1},\mathscr{S})^a, d(y_0,\mathscr{S})^a\} \\ &\le \varepsilon^{6/\beta}e^{a\varepsilon}\min\{d(x_0,\mathscr{S})^a, d(x_1,\mathscr{S})^a\} < \varepsilon\mathfrak{r}(x_0). \end{aligned}$$

Also, (A6) implies that $d(x_1,y_0) < \rho(\widehat{x})^{-a}d(x_0,y_{-1}) < \rho(\widehat{x})^{-a}Q(\widehat{x}) \ll \varepsilon\rho(\widehat{x})^a < \varepsilon\mathfrak{r}(x_0)$ and so $A \subset B(y_0, 40Q(\widehat{y})) \subset B(x_1, 40Q(\widehat{y}) + d(x_1,y_0)) \subset E_{x_0}$. Now we prove the second inclusion. Since $A \subset B(y_0, 40Q(\widehat{y}))$, again by (A6) we have

$$\begin{aligned} f^{-1}_{\widehat{f}^{-1}(\widehat{y})}(A) &\subset B(y_{-1}, 40d(y_{-1},\mathscr{S})^{-a}Q(\widehat{y})) \\ &\subset B(x_0, 40d(y_{-1},\mathscr{S})^{-a}Q(\widehat{y}) + d(y_{-1},x_0)). \end{aligned}$$

Assumption (A6) also implies that $f^{-1}_{\widehat{x}}(E_{x_0}) \supset B(x_0, 2d(x_0,\mathscr{S})^a\mathfrak{r}(x_0))$. Since

$$\begin{aligned} &40d(y_{-1},\mathscr{S})^{-a}Q(\widehat{y}) + d(y_{-1},x_0) \\ &\quad< \min\{d(y_{-1},\mathscr{S})^{3a}, d(y_0,\mathscr{S})^{3a}\} \\ &\quad\le e^{3a\varepsilon}\min\{d(x_0,\mathscr{S})^{3a}, d(x_1,\mathscr{S})^{3a}\} \\ &\quad< 2d(x_0,\mathscr{S})^a\mathfrak{r}(x_0), \end{aligned}$$

we obtain that $f^{-1}_{\widehat{f}^{-1}(\widehat{y})}(A) \subset f^{-1}_{\widehat{x}}(E_{x_0})$. □

Now we prove a version of Theorem 3.3 for $F^{\pm}_{\widehat{x},\widehat{y}}$.

THEOREM 4.3. *The following holds for all $\varepsilon > 0$ small enough. If $\widehat{x},\widehat{y} \in \mathrm{NUH}$ and $\Psi^{\eta}_{\widehat{f}(\widehat{x})} \overset{\varepsilon}{\approx} \Psi^{\eta'}_{\widehat{y}}$ then $F_{\widehat{x},\widehat{y}}$ is well-defined in $B[20Q(\widehat{x})]$ and can be written as $F_{\widehat{x},\widehat{y}} = D(\widehat{x}) + H^+$ s.t. for all $t \in [\eta, 20Q(\widehat{x})]$ it holds:*

(a) $\|H^+(0)\| < \varepsilon\eta$.

(b) $\|H^+\|_{C^1(B[t])} < \varepsilon t^{\beta/3}$.
(c) $\mathrm{Hol}_{\frac{\beta}{3}}(dH^+) < \varepsilon$.

Similarly, if $\widehat{x},\widehat{y} \in \mathrm{NUH}$ *and* $\Psi^{\eta}_{\widehat{x}} \overset{\varepsilon}{\approx} \Psi^{\eta'}_{\widehat{f}^{-1}(\widehat{y})}$ *then* $F^{-1}_{\widehat{x},\widehat{y}}$ *is well-defined in* $B[20Q(\widehat{y})]$ *and can be written as* $F^{-1}_{\widehat{x},\widehat{y}} = D(\widehat{f}^{-1}(\widehat{y}))^{-1} + H^-$ *s.t. for all* $t \in [\eta, 20Q(\widehat{y})]$ *it holds:*

(d) $\|H^-(0)\| < \varepsilon\eta$.
(e) $\|H^-\|_{C^1(B[t])} < \varepsilon t^{\beta/3}$.
(f) $\mathrm{Hol}_{\frac{\beta}{3}}(dH^-) < \varepsilon$.

Observe that the overlap condition is strong enough to guarantee that $F^{\pm}_{\widehat{x},\widehat{y}}$ is defined in a domain of radius $20Q(\cdot)$ and, except for the C^1 norm, the estimates are also uniform in this domain.

PROOF. First assume that $\Psi^{\eta}_{\widehat{f}(\widehat{x})} \overset{\varepsilon}{\approx} \Psi^{\eta'}_{\widehat{y}}$, and let us prove items (a)–(c). We write $F_{\widehat{x},\widehat{y}} = (\Psi^{-1}_{\widehat{y}} \circ \Psi_{\widehat{f}(\widehat{x})}) \circ F_{\widehat{x}} =: G \circ F_{\widehat{x}}$ and see it as a small perturbation of $F_{\widehat{x}}$. By Theorem 3.3(3–4),

$$F_{\widehat{x}}(0) = 0, \|dF_{\widehat{x}}\|_{C^0} < 2d(\vartheta[\widehat{x}], \mathscr{S})^{-a}(1+e^{2\chi})^{1/2}, \|d(F_{\widehat{x}})_v - d(F_{\widehat{x}})_w\| \leq \varepsilon\|v-w\|^{\beta/2}$$

for $v, w \in B[20Q(\widehat{x})]$, where the C^0 norm is taken in $B[20Q(\widehat{x})]$, and by Proposition 4.1(5) we have

$$\|G - \mathrm{Id}\| < \varepsilon(\eta\eta')^2,\ \|d(G-\mathrm{Id})\|_{C^0} < \varepsilon(\eta\eta')^2,\ \|dG_v - dG_w\| \leq \varepsilon(\eta\eta')^2\|v-w\|^{\beta/2}$$

for $v, w \in B[d(\vartheta_1[\widehat{x}], \mathscr{S})^{2a}]$, where the C^0 norm is taken in this same domain. We start proving that $F_{\widehat{x},\widehat{y}}$ is well-defined in $B[20Q(\widehat{x})]$. We have

$$F_{\widehat{x}}(B[20Q(\widehat{x})]) \subset B[40\rho(\widehat{x})^{-a}(1+e^{2\chi})^{1/2}Q(\widehat{x})] \subset B[d(\vartheta_1[\widehat{x}], \mathscr{S})^{2a}]$$

since $40\rho(\widehat{x})^{-a}(1+e^{2\chi})^{1/2}Q(\widehat{x}) < 40(1+e^{2\chi})^{1/2}\varepsilon^{6/\beta}d(\vartheta_1[\widehat{x}], \mathscr{S})^{2a} < d(\vartheta_1[\widehat{x}], \mathscr{S})^{2a}$ for $\varepsilon > 0$ small enough. By Proposition 4.1(5), $F_{\widehat{x},\widehat{y}}$ is well-defined.

Let $H^+ := G \circ F_{\widehat{x}} - d(F_{\widehat{x}})_0 = F_{\widehat{x},\widehat{y}} - D(\widehat{x})$, cf. Lemma 3.1. By Proposition 4.1(3), we have $\|H^+(0)\| = \|G(0)\| < \varepsilon(\eta\eta')^2 < \frac{1}{8}\varepsilon\eta$, which is stronger than (a). Now we prove (c). Start observing that for $\varepsilon > 0$ small enough it holds:

$$\begin{aligned}
\|d(H^+)_0\| &\leq \|dG_0 \circ D(\widehat{x}) - D(\widehat{x})\| \\
&\leq \|d(G-\mathrm{Id})_0\| \cdot \|D(\widehat{x})\| \\
&< \varepsilon(\eta\eta')^2 d(\vartheta[\widehat{x}], \mathscr{S})^{-a}(1+e^{2\chi})^{1/2} \\
&< \varepsilon\eta\eta'\varepsilon^{6/\beta}(1+e^{2\chi})^{1/2} \\
&< \frac{1}{8}\varepsilon\eta^{\beta/3}.
\end{aligned}$$

Since $F_{\widehat{x}}(B[20Q(\widehat{x})]) \subset B[d(\vartheta_1[\widehat{x}],\mathscr{S})^{2a}]$, if $\varepsilon > 0$ is small enough then for all $v, w \in B[20Q(\widehat{x})]$ we have

$$\begin{aligned}
&\|d(H^+)_v - d(H^+)_w\| \\
&= \|dG_{F_{\widehat{x}}(v)} \circ d(F_{\widehat{x}})_v - dG_{F_{\widehat{x}}(w)} \circ d(F_{\widehat{x}})_w\| \\
&\le \|dG_{F_{\widehat{x}}(v)} - dG_{F_{\widehat{x}}(w)}\| \cdot \|d(F_{\widehat{x}})_v\| + \|dG_{F_{\widehat{x}}(w)}\| \cdot \|d(F_{\widehat{x}})_v - d(F_{\widehat{x}})_w\| \\
&\le \varepsilon(\eta\eta')^2 \|F_{\widehat{x}}(v) - F_{\widehat{x}}(w)\|^{\beta/2} \|d(F_{\widehat{x}})\|_{C^0} + \varepsilon \|dG\|_{C^0} \|v-w\|^{\beta/2} \\
&\le \left[\varepsilon(\eta\eta')^2 \|d(F_{\widehat{x}})\|_{C^0}^{1+\beta/2} + 80\varepsilon Q(\widehat{x})^{\beta/6}\right] \|v-w\|^{\beta/3} \\
&\le \left[4\eta d(\vartheta[\widehat{x}],\mathscr{S})^{-2a}(1+e^{2\chi}) + 80Q(\widehat{x})^{\beta/6}\right] \varepsilon\|v-w\|^{\beta/3} \\
&\le \left[4\varepsilon^{6/\beta}(1+e^{2\chi}) + 80\varepsilon\right] \varepsilon\|v-w\|^{\beta/3} < \tfrac{1}{8}\varepsilon\|v-w\|^{\beta/3},
\end{aligned}$$

hence $\mathrm{Hol}_{\frac{\beta}{3}}(dH^+) < \frac{1}{8}\varepsilon$, which is stronger than (c). Finally, we fix $\eta \le t \le 20Q(\widehat{x})$ and prove (b). For $v \in B[t]$:

- $\|d(H^+)_v\| \le \|d(H^+)_0\| + \frac{1}{8}\varepsilon\|v\|^{\beta/3} < \frac{1}{8}\varepsilon\eta^{\beta/3} + \frac{1}{8}\varepsilon t^{\beta/3} \le \frac{1}{4}\varepsilon t^{\beta/3}$, therefore $\|dH^+\|_{C^0(B[t])} < \frac{1}{4}\varepsilon t^{\beta/3}$.
- $\|H^+(v)\| \le \|H^+(0)\| + \|dH^+\|_{C^0}\|v\| < \frac{1}{8}\varepsilon\eta + \frac{1}{4}\varepsilon t^{\beta/3} < \frac{1}{2}\varepsilon t^{\beta/3}$, which implies that $\|H^+\|_{C^0(B[t])} < \frac{1}{2}\varepsilon t^{\beta/3}$.

Thus, $\|H^+\|_{C^1(B[t])} < \varepsilon t^{\beta/3}$.

Now we prove the second part. Assume that $\Psi_{\widehat{x}}^{\eta} \overset{\varepsilon}{\approx} \Psi_{\widehat{f}^{-1}(\widehat{y})}^{\eta'}$. By Lemma 4.2, we have $F_{\widehat{x},\widehat{y}}^{-1} = \Psi_{\widehat{x}}^{-1} \circ f_{\widehat{f}^{-1}(\widehat{y})}^{-1} \circ \Psi_{\widehat{y}}$ in the domain $B[20Q(\widehat{y})]$. Writing $F_{\widehat{x},\widehat{y}}^{-1} = (\Psi_{\widehat{x}}^{-1} \circ \Psi_{\widehat{f}^{-1}(\widehat{y})}) \circ F_{\widehat{f}^{-1}(\widehat{y})}^{-1} =: G \circ L$ and seeing it as a small perturbation of $L = F_{\widehat{f}^{-1}(\widehat{y})}^{-1}$, we have by Theorem 3.3(3–4) that

$$L(0) = 0,\ \|dL\|_{C^0} < 2d(\vartheta_{-1}[\widehat{y}],\mathscr{S})^{-a}(1+e^{2\chi})^{1/2},\ \|dL_v - dL_w\| \le \varepsilon\|v-w\|^{\beta/2}$$

for $v, w \in B[20Q(\widehat{y})]$, where the C^0 norm is taken in $B[20Q(\widehat{y})]$, and by Proposition 4.1(5) that

$$\|G - \mathrm{Id}\| < \varepsilon(\eta\eta')^2,\ \|d(G-\mathrm{Id})\|_{C^0} < \varepsilon(\eta\eta')^2,\ \|dG_v - dG_w\| \le \varepsilon(\eta\eta')^2\|v-w\|^{\beta/2}$$

for $v, w \in B[d(y_{-1},\mathscr{S})^{2a}]$, where the C^0 norm is taken in this same domain. To prove that $F_{\widehat{x},\widehat{y}}$ is well-defined in $B[20Q(\widehat{y})]$, notice that

$$L(B[20Q(\widehat{y})]) \subset B[40\rho(\widehat{y})^{-a}(1+e^{2\chi})^{1/2}Q(\widehat{y})] \subset B[d(y_{-1},\mathscr{S})^{2a}]$$

since $40\rho(\widehat{y})^{-a}(1+e^{2\chi})^{1/2}Q(\widehat{y}) < 40(1+e^{2\chi})^{1/2}\varepsilon^{6/\beta}d(y_{-1},\mathscr{S})^{2a} < d(y_{-1},\mathscr{S})^{2a}$ for $\varepsilon > 0$ small enough. By Proposition 4.1(5), $F_{\widehat{x},\widehat{y}}^{-1}$ is well-defined.

Let $H^- := G \circ L - dL_0 = F_{\widehat{x},\widehat{y}}^{-1} - D(\widehat{f}^{-1}(\widehat{y}))^{-1}$. By Proposition 4.1(5), we have $\|H^-(0)\| = \|G(0)\| < \varepsilon(\eta\eta')^2 < \frac{1}{8}\varepsilon\eta$, which is stronger than (d). To prove (f), start observing that for $\varepsilon > 0$ small enough it holds:

$$\begin{aligned}
&\|d(H^-)_0\| \le \|dG_0 \circ D(\widehat{f}^{-1}(\widehat{y}))^{-1} - D(\widehat{f}^{-1}(\widehat{y}))^{-1}\| \le \|d(G-\mathrm{Id})_0\|\|D(\widehat{f}^{-1}(\widehat{y}))^{-1}\| \\
&< \varepsilon(\eta\eta')^2 d(\vartheta_{-1}[\widehat{y}],\mathscr{S})^{-a}(1+e^{2\chi})^{1/2} < \varepsilon\eta\eta'\varepsilon^{6/\beta}(1+e^{2\chi})^{1/2} < \tfrac{1}{8}\varepsilon\eta^{\beta/3}.
\end{aligned}$$

Since $L(B[20Q(\widehat{y})]) \subset B[d(y_{-1},\mathscr{S})^{2a}]$, if $\varepsilon>0$ is small enough then for all $v,w\in B[20Q(\widehat{y})]$ we have:

$$\begin{aligned}\|d(H^-)_v-d(H^-)_w\| &= \|dG_{L(v)}\circ dL_v - dG_{L(w)}\circ dL_w\|\\ &\leq \left\|dG_{L(v)}-dG_{L(w)}\right\|\cdot\|dL_v\| + \left\|dG_{L(w)}\right\|\cdot\|dL_v-dL_w\|\\ &\leq \varepsilon(\eta\eta')^2\|L(v)-L(w)\|^{\beta/2}\|dL\|_{C^0}+\varepsilon\|dG\|_{C^0}\|v-w\|^{\beta/2}\\ &\leq \left[\varepsilon(\eta\eta')^2\|dL\|_{C^0}^{1+\beta/2}+80\varepsilon Q(\widehat{y})^{\beta/6}\right]\|v-w\|^{\beta/3}\\ &\leq \left[4\eta' d(\vartheta_{-1}[\widehat{y}],\mathscr{S})^{-2a}(1+e^{2\chi})+80Q(\widehat{y})^{\beta/6}\right]\varepsilon\|v-w\|^{\beta/3}\\ &\leq \left[4\varepsilon^{6/\beta}(1+e^{2\chi})+80\varepsilon\right]\varepsilon\|v-w\|^{\beta/3}<\tfrac{1}{8}\varepsilon\|v-w\|^{\beta/3}.\end{aligned}$$

This implies that $\mathrm{Hol}_{\frac{\beta}{3}}(dH^-)<\frac{1}{8}\varepsilon$, which is stronger than (f). Finally, we fix $\eta\leq t\leq 20Q(\widehat{y})$ and prove (e). For $v\in B[t]$:

- $\|d(H^-)_v\|\leq\|d(H^-)_0\|+\frac{1}{8}\varepsilon\|v\|^{\beta/3}<\frac{1}{8}\varepsilon\eta^{\beta/3}+\frac{1}{8}\varepsilon t^{\beta/3}\leq\frac{1}{4}\varepsilon t^{\beta/3}$, therefore $\|dH^-\|_{C^0(B[t])}<\frac{1}{4}\varepsilon t^{\beta/3}$.
- $\|H^-(v)\|\leq\|H^-(0)\|+\|dH^-\|_{C^0}\|v\|<\frac{1}{8}\varepsilon\eta+\frac{1}{4}\varepsilon t^{\beta/3}<\frac{1}{2}\varepsilon t^{\beta/3}$, which implies that $\|H^-\|_{C^0(B[t])}<\frac{1}{2}\varepsilon t^{\beta/3}$.

Thus, $\|H^-\|_{C^1(B[t])}<\varepsilon t^{\beta/3}$. □

4.3. Double charts

We now define ε-double charts. Recall that $\varepsilon>0$ is a fixed small parameter, and that $\delta_\varepsilon=e^{-\varepsilon n}\in I_\varepsilon$ where n is the unique positive integer s.t. $e^{-\varepsilon n}<\varepsilon\leq e^{-\varepsilon(n-1)}$.

ε-DOUBLE CHART: An *ε-double chart* is a pair of Pesin charts $\Psi_{\widehat{x}}^{p^s,p^u}=(\Psi_{\widehat{x}}^{p^s},\Psi_{\widehat{x}}^{p^u})$ where $p^s,p^u\in I_\varepsilon$ with $0<p^s,p^u\leq\delta_\varepsilon Q(\widehat{x})$.

When $v=\Psi_{\widehat{x}}^{p^s,p^u}$, we call $\widehat{x}$ the center of v. The parameters p^s/p^u will define sizes for the stable/unstable manifold at $\widehat{x}$. Hence, for $\widehat{x}\in\mathrm{NUH}^*$ the prototype of ε-double chart is $\Psi_{\widehat{x}}^{q^s(\widehat{x}),q^u(\widehat{x})}$.

EDGE $v\overset{\varepsilon}{\to}w$: Given ε-double charts $v=\Psi_{\widehat{x}}^{p^s,p^u}$ and $w=\Psi_{\widehat{y}}^{q^s,q^u}$, we draw an edge from v to w if the following conditions are satisfied:

(GPO1) $\Psi_{\widehat{f}(\widehat{x})}^{q^s\wedge q^u}\overset{\varepsilon}{\approx}\Psi_{\widehat{y}}^{q^s\wedge q^u}$ and $\Psi_{\widehat{f}^{-1}(\widehat{y})}^{p^s\wedge p^u}\overset{\varepsilon}{\approx}\Psi_{\widehat{x}}^{p^s\wedge p^u}$.

(GPO2) $p^s=\min\{e^\varepsilon q^s,\delta_\varepsilon Q(\widehat{x})\}$ and $q^u=\min\{e^\varepsilon p^u,\delta_\varepsilon Q(\widehat{y})\}$.

Since $d_{s/u}$ are $\widehat{f}$-invariant, (GPO1) implies that $d_{s/u}(\widehat{x})=d_{s/u}(\widehat{y})$. Condition (GPO1) allows to pass from an ε-double chart at $\widehat{x}$ to an ε-double chart at $\widehat{y}$ and vice-versa. By Lemma 3.5(2), we have $\Psi_{\widehat{x}}^{q^s(\widehat{x}),q^u(\widehat{x})}\overset{\varepsilon}{\to}\Psi_{\widehat{f}(\widehat{x})}^{q^s(\widehat{f}(\widehat{x})),q^u(\widehat{f}(\widehat{x}))}$. Hence we can regard (GPO2) as the counterpart of Lemma 3.5(2) for a general edge. Observe that (GPO2) implies that $\frac{p^s\wedge p^u}{q^s\wedge q^u}=e^{\pm\varepsilon}$. Indeed, we have $p^s\wedge p^u=\min\{e^\varepsilon q^s,\delta_\varepsilon Q(\widehat{x})\}\wedge p^u=\min\{e^\varepsilon q^s,p^u\}$ and similarly $q^s\wedge q^u=\min\{q^s,e^\varepsilon p^u\}$. Thus $p^s\wedge p^u=e^\varepsilon\min\{q^s,e^{-\varepsilon}p^u\}\leq e^\varepsilon\min\{q^s,e^\varepsilon p^u\}=e^\varepsilon(q_s\wedge q_u)$ and similarly $q_s\wedge q_u\leq e^\varepsilon(p_s\wedge p_u)$.

ε-GENERALIZED PSEUDO-ORBIT (ε-GPO): An *ε-generalized pseudo-orbit (ε-gpo)* is a sequence $\underline{v}=\{v_n\}_{n\in\mathbb{Z}}$ of ε-double charts s.t. $v_n\overset{\varepsilon}{\to}v_{n+1}$ for all $n\in\mathbb{Z}$.

4.4. The graph transform method

The proofs of this section are collected, in a self-contained manner, in the Appendix. We hope this separation will both clarify the presentation as highlight the main arguments used to employ the graph transform method, without boundedness assumptions on the derivative. Let $v = \Psi_{\widehat{x}}^{p^s,p^u}$ be an ε-double chart. Recall from Lemma 3.1 that

$$D(\widehat{x}) = \begin{bmatrix} D_s(\widehat{x}) & 0 \\ 0 & D_u(\widehat{x}) \end{bmatrix}$$

where $D_s(\widehat{x})$ is a $d_s(\widehat{x}) \times d_s(\widehat{x})$ matrix s.t. $\|D_s(\widehat{x})\| < e^{-\chi}$, and $D_u(\widehat{x})$ is a $d_u(\widehat{x}) \times d_u(\widehat{x})$ matrix s.t. $\|D_u(\widehat{x})^{-1}\| < e^{-\chi}$.

ADMISSIBLE MANIFOLDS: We define an *s-admissible manifold at v* as a set of the form $\Psi_{\widehat{x}}\{(v_1, G(v_1)) : v_1 \in B^{d_s(\widehat{x})}[p^s]\}$ where $G : B^{d_s(\widehat{x})}[p^s] \to \mathbb{R}^{d_u(\widehat{x})}$ is a $C^{1+\beta/3}$ function s.t.:

(AM1) $\|G(0)\| \leq 10^{-3}(p^s \wedge p^u)$.
(AM2) $\|dG_0\| \leq \frac{1}{2}(p^s \wedge p^u)^{\beta/3}$.
(AM3) $\|dG\|_{C^0} + \mathrm{Hol}_{\frac{\beta}{3}}(dG) \leq \frac{1}{2}$ where the norms are taken in $B^{d_s(\widehat{x})}[p^s]$.

Similarly, a *u-admissible manifold at v* is a set of the form $\Psi_{\widehat{x}}\{(G(v_2), v_2) : v_2 \in B^{d_u(\widehat{x})}[p^u]\}$ where $G : B^{d_u(\widehat{x})}[p^u] \to \mathbb{R}^{d_s(\widehat{x})}$ is a $C^{1+\beta/3}$ function satisfying (AM1)–(AM3), where the norms are taken in $B^{d_u(\widehat{x})}[p^u]$.

The function G is called the *representing function*. Since we are working with the euclidean norm, admissible manifolds at v are contained in $\Psi_{\widehat{x}}(B^{d_s(\widehat{x})}[p^s] \times B^{d_u(\widehat{x})}[p^s])$. We let $\mathscr{M}^s(v)$ (resp. $\mathscr{M}^u(v)$) denote the set of all s-admissible (resp. u-admissible) manifolds at v. Define a metric on $\mathscr{M}^s(v)$, comparing representing functions: for $V_1, V_2 \in \mathscr{M}^s(v)$ with representing functions G_1, G_2 and for $i \geq 0$, define $d_{C^i}(V_1, V_2) := \|G_1 - G_2\|_{C^i}$ where the norm is taken in $B^{d_s(\widehat{x})}[p^s]$. A similar definition holds for $\mathscr{M}^u(v)$.

LEMMA 4.4. *The following holds for $\varepsilon > 0$ small enough. If $v = \Psi_{\widehat{x}}^{p^s,p^u}$ is an ε-double chart, then for every $V^s \in \mathscr{M}^s(v)$ and $V^u \in \mathscr{M}^u(v)$ the intersection $V^s \cap V^u$ consists of a single point $y = \Psi_{\widehat{x}}(w)$, and $\|w\| < 50^{-1}(p^s \wedge p^u)$. Additionally, y is a Lipschitz function of the pair (V^s, V^u), with Lipschitz constant* 3.

Let $v = \Psi_{\widehat{x}}^{p^s,p^u}$, $w = \Psi_{\widehat{y}}^{q^s,q^u}$ be ε-double charts with $v \xrightarrow{\varepsilon} w$. We now define the *graph transforms*: these are two maps that work in different directions of the edge $v \xrightarrow{\varepsilon} w$, one of them sending u-admissible manifolds at v to u-admissible manifolds at w, and the other sending s-admissible manifolds at w to s-admissible manifolds at v. In the sequel, we will write $d = d_s(\widehat{x}) = d_s(\widehat{y})$ and $m - d = d_u(\widehat{x}) = d_u(\widehat{y})$.

GRAPH TRANSFORMS $\mathscr{F}_{v,w}^s$ AND $\mathscr{F}_{v,w}^u$: The *graph transform* $\mathscr{F}_{v,w}^s : \mathscr{M}^s(w) \to \mathscr{M}^s(v)$ is the map that sends an s-admissible manifold at w with representing function $G : B^d[q^s] \to \mathbb{R}^{m-d}$ to the unique s-admissible manifold at v with representing function $\widetilde{G} : B^d[p^s] \to \mathbb{R}^{m-d}$ s.t.

$$\left\{(v_1, \widetilde{G}(v_1)) : v_1 \in B^d[p^s]\right\} \subset F_{\widehat{x},\widehat{y}}^{-1}\left\{(v_1, G(v_1)) : v_1 \in B^d[q^s]\right\}.$$

The *graph transform* $\mathscr{F}^u_{v,w}:\mathscr{M}^u(v)\to\mathscr{M}^u(w)$ is the map that sends a u-admissible manifold at v with representing function $G: B^{m-d}[p^u]\to\mathbb{R}^d$ to the unique u-admissible manifold at w with representing function $\widetilde{G}: B^{m-d}[q^u]\to\mathbb{R}^d$ s.t.

$$\left\{(\widetilde{G}(v_2),v_2): v_2\in B^{m-d}[q^u]\right\}\subset F_{\widehat{x},\widehat{y}}\left\{(G(v_2),v_2): v_2\in B^{m-d}[p^u]\right\}.$$

In other words, the representing functions of s,u-admissible manifolds change by the application of $F^{-1}_{\widehat{x},\widehat{y}}, F_{\widehat{x},\widehat{y}}$ respectively.

PROPOSITION 4.5. *The following holds for $\varepsilon>0$ small enough. If $v\overset{\varepsilon}{\to}w$ then $\mathscr{F}^s_{v,w}$ and $\mathscr{F}^u_{v,w}$ are well-defined. Furthermore, if $V_1,V_2\in\mathscr{M}^u(v)$ then:*

(1) $d_{C^0}(\mathscr{F}^u_{v,w}(V_1),\mathscr{F}^u_{v,w}(V_2))\le e^{-\chi/2}d_{C^0}(V_1,V_2)$.
(2) $d_{C^1}(\mathscr{F}^u_{v,w}(V_1),\mathscr{F}^u_{v,w}(V_2))\le e^{-\chi/2}(d_{C^1}(V_1,V_2)+d_{C^0}(V_1,V_2)^{\beta/3})$.
(3) *$f(V_i)$ intersects every element of $\mathscr{M}^s(w)$ at exactly one point.*

An analogous statement holds for $\mathscr{F}^s_{v,w}$, where in (3) we consider $f^{-1}_{\widehat{x}}(V_i)$.

As already mentioned, all the proofs are in the Appendix, where the required assumptions (GT1)–(GT3) are weaker than we need. If $\Psi^{p^s,p^u}_{\widehat{x}}\overset{\varepsilon}{\to}\Psi^{q^s,q^u}_{\widehat{y}}$ then (GT1)–(GT3) are satisfied for:

- $p=q^s,\widetilde{p}=p^s,\eta=q^s\wedge q^u,\widetilde{\eta}=p^s\wedge p^u$ and $F=F^{-1}_{\widehat{x},\widehat{y}}$, thus $\mathscr{F}^s_{v,w}$ satisfies the appendix.
- $p=p^u,\widetilde{p}=q^u,\eta=p^s\wedge p^u,\widetilde{\eta}=q^s\wedge q^u$ and $F=F_{\widehat{x},\widehat{y}}$, thus $\mathscr{F}^u_{v,w}$ satisfies the appendix.

4.5. Stable/unstable manifolds of ε-gpo's

Call a sequence $\underline{v}^+=\{v_n\}_{n\ge0}$ a *positive ε-gpo* if $v_n\overset{\varepsilon}{\to}v_{n+1}$ for all $n\ge0$. Similarly, a *negative ε-gpo* is a sequence $\underline{v}^-=\{v_n\}_{n\le0}$ s.t. $v_{n-1}\overset{\varepsilon}{\to}v_n$ for all $n\le0$.

STABLE/UNSTABLE MANIFOLD OF POSITIVE/NEGATIVE ε-GPO: The *stable manifold* of a positive ε-gpo $\underline{v}^+=\{v_n\}_{n\ge0}$ is

$$V^s[\underline{v}^+]:=\lim_{n\to\infty}(\mathscr{F}^s_{v_0,v_1}\circ\cdots\circ\mathscr{F}^s_{v_{n-2},v_{n-1}}\circ\mathscr{F}^s_{v_{n-1},v_n})(V_n)$$

for some (any) choice of $(V_n)_{n\ge0}$ with $V_n\in\mathscr{M}^s(v_n)$. The *unstable manifold* of a negative ε-gpo $\underline{v}^-=\{v_n\}_{n\le0}$ is

$$V^u[\underline{v}^-]:=\lim_{n\to-\infty}(\mathscr{F}^u_{v_{-1},v_0}\circ\cdots\circ\mathscr{F}^u_{v_{n+1},v_{n+2}}\circ\mathscr{F}^u_{v_n,v_{n+1}})(V_n)$$

for some (any) choice of $(V_n)_{n\le0}$ with $V_n\in\mathscr{M}^u(v_n)$. For an ε-gpo $\underline{v}=\{v_n\}_{n\in\mathbb{Z}}$, let $V^s[\underline{v}]:=V^s[\{v_n\}_{n\ge0}]$ and $V^u[\underline{v}]:=V^u[\{v_n\}_{n\le0}]$.

Before proceeding to the main properties of these manifolds, let us show that the action of the graph transforms has a simple description in terms of positive/negative ε-gpo's.

LEMMA 4.6. *Let $v\overset{\varepsilon}{\to}w$. If $\underline{v}^-=\{v_n\}_{n\le0}$ is a negative ε-gpo with $v_0=v$, then*

$$\mathscr{F}^u_{v,w}(V^u[\underline{v}^-])=V^u[\underline{w}^-]$$

where $\underline{w}^-=\{w_n\}_{n\le0}$ is the negative ε-gpo with $w_0=w$ and $w_n=v_{n+1}$ for $n<0$. Similarly, if $\underline{w}^+=\{w_n\}_{n\ge0}$ is a positive ε-gpo with $w_0=w$, then

$$\mathscr{F}^s_{v,w}(V^s[\underline{w}^+])=V^s[\underline{v}^+]$$

where $\underline{v}^+ = \{v_n\}_{n\geq 0}$ *is the positive* ε*-gpo with* $v_0 = v$ *and* $v_n = w_{n-1}$ *for* $n > 0$.

PROOF. We prove the first statement. By definition,

$$\begin{aligned} V^u[\underline{w}^-] &= \lim_{n\to-\infty}(\mathscr{F}^u_{w_{-1},w_0} \circ \mathscr{F}^u_{w_{-2},w_{-1}} \circ \cdots \circ \mathscr{F}^u_{w_n,w_{n+1}})(V_n) \\ &= \lim_{n\to-\infty}(\mathscr{F}^u_{v,w} \circ \mathscr{F}^u_{v_{-1},v_0} \circ \cdots \circ \mathscr{F}^u_{v_{n+1},v_{n+2}})(V_n). \end{aligned}$$

for some (any) choice of $(V_n)_{n\leq 0}$ with $V_n \in \mathscr{M}^u(w_n) = \mathscr{M}^u(v_{n+1})$. By Proposition 4.5, $\mathscr{F}^u_{v,w}$ is continuous and so

$$V^u[\underline{w}^-] = \mathscr{F}^u_{v,w}\left[\lim_{n\to-\infty}(\mathscr{F}^u_{v_{-1},v_0} \circ \cdots \circ \mathscr{F}^u_{v_{n+1},v_{n+2}})(V_n)\right] = \mathscr{F}^u_{v,w}(V^u[\underline{v}^-]).$$

The second statement is proved analogously. □

The next proposition collects the main properties of stable/unstable manifolds.

PROPOSITION 4.7. *The following holds for all* $\varepsilon > 0$ *small enough. For every positive* ε*-gpo* $\underline{v}^+ = \{\Psi^{p^s_n,p^u_n}_{\widehat{x}_n}\}_{n\geq 0}$ *and every negative* ε*-gpo* $\underline{v}^- = \{\Psi^{p^s_n,p^u_n}_{\widehat{x}_n}\}_{n\leq 0}$*:*

(1) ADMISSIBILITY: *$V^s[\underline{v}^+]$ is a well-defined admissible manifolds at v_0, and if G and G_n are the representing functions of $V^s[\underline{v}^+]$ and $(\mathscr{F}^s_{v_0,v_1} \circ \cdots \circ \mathscr{F}^s_{v_{n-1},v_n})(V_n)$ respectively, then $dG_n \to dG$ uniformly. The same holds for $V^u[\underline{v}^-]$.*

(2) INVARIANCE:

$$f(V^s[\{v_n\}_{n\geq 0}]) \subset V^s[\{v_n\}_{n\geq 1}] \text{ and } f^{-1}_{\widehat{x}_{-1}}(V^u[\{v_n\}_{n\leq 0}]) \subset V^u[\{v_n\}_{n\leq -1}].$$

(3) SHADOWING: *If $d = d_s(\widehat{x}_0)$ then $V^s[\underline{v}^+]$ equals*

$$\left\{y \in \Psi_{\widehat{x}_0}(B^d[p^s_0] \times B^{m-d}[p^s_0]) : \begin{array}{c} f^n(y) \in \Psi_{\widehat{x}_n}(B[20Q(\widehat{x}_n)]), \\ \forall n \geq 0 \end{array}\right\}$$

and $V^u[\underline{v}^-]$ equals

$$\left\{y \in \Psi_{\widehat{x}_0}(B^d[p^u_0] \times B^{m-d}[p^u_0]) : \begin{array}{c} (f^{-1}_{\widehat{x}_n} \circ \cdots \circ f^{-1}_{\widehat{x}_{-1}})(y) \in \Psi_{\widehat{x}_n}(B[20Q(\widehat{x}_n)]), \\ \forall n < 0 \end{array}\right\}.$$

(4) HYPERBOLICITY: *If $y, z \in V^s[\underline{v}^+]$ then*

$$d(f^n(y), f^n(z)) \leq 4p^s_0 e^{-\frac{\chi}{2}n} \text{ for all } n \geq 0.$$

Similarly, if $y, z \in V^u[\underline{v}^-]$ then

$$d((f^{-1}_{\widehat{x}_n} \circ \cdots \circ f^{-1}_{\widehat{x}_{-1}})(y), (f^{-1}_{\widehat{x}_n} \circ \cdots \circ f^{-1}_{\widehat{x}_{-1}})(z)) \leq 4p^u_0 e^{\frac{\chi}{2}n} \text{ for all } n \leq 0.$$

(5) HÖLDER PROPERTY: *The map $\underline{v}^+ \mapsto V^s[\underline{v}^+]$ is Hölder continuous, i.e. there exists $K > 0$ and $\theta < 1$ s.t. for all $N \geq 0$, if $\underline{v}^+, \underline{w}^+$ are positive ε-gpo's with $v_n = w_n$ for $n = 0, \ldots, N$ then $d_{C^1}(V^s[\underline{v}^+], V^s[\underline{w}^+]) \leq K\theta^N$. The same holds for the map $\underline{v}^- \mapsto V^u[\underline{v}^-]$.*

PROOF. When M is compact and f is a $C^{1+\beta}$ diffeomorphism, this is [**BO18**, Proposition 3.12 and 4.4(1)]. The same proofs work in our case, since they only use the hyperbolicity of $F^{\pm 1}_{\widehat{x}_n,\widehat{x}_{n+1}}$ (Theorem 4.3) and the contracting properties of the graph transforms (Proposition 4.5). □

4.6. Stable/unstable sets of ε-gpo's

Having in mind that we want to build a symbolic dynamics for $\widehat{f}:\widehat{M}\to\widehat{M}$, we now introduce the set-theoretical counterparts of stable/unstable manifolds in $\widehat{M}$, which we call stable/unstable sets.

STABLE/UNSTABLE SET OF POSITIVE/NEGATIVE ε-GPO: The *stable set* of a positive ε-gpo $\underline{v}^+ = \{\Psi_{\widehat{x}_n}^{p_n^s,p_n^u}\}_{n\geq 0}$ is

$$\widehat{V}^s[\underline{v}^+] := \{\widehat{y}=(y_n)_{n\in\mathbb{Z}}\in\widehat{M} : y_0\in V^s[\underline{v}^+]\}.$$

The *unstable set* of a negative ε-gpo $\underline{v}^- = \{\Psi_{\widehat{x}_n}^{p_n^s,p_n^u}\}_{n\leq 0}$ is

$$\widehat{V}^u[\underline{v}^-] := \{\widehat{y}=(y_n)_{n\in\mathbb{Z}}\in\widehat{M} : y_0\in V^u[\underline{v}^-] \text{ and } y_n = f_{\widehat{x}_n}^{-1}(y_{n+1}) \text{ for all } n<0\}.$$

For an ε-gpo $\underline{v}=\{v_n\}_{n\in\mathbb{Z}}$, let $\widehat{V}^s[\underline{v}] := \widehat{V}^s[\{v_n\}_{n\geq 0}]$ and $\widehat{V}^u[\underline{v}] := \widehat{V}^u[\{v_n\}_{n\leq 0}]$.

In other words, $\widehat{V}^s[\underline{v}^+]$ is the set of all points in $\widehat{M}$ for which the zeroth position belongs to $V^s[v^+]$, while $\widehat{V}^u[\underline{v}^-]$ is the set of all points in $\widehat{M}$ for which the zeroth position belongs to $V^u[v^-]$ and *additionally* the negative positions are uniquely determined by the inverse branches prescribed by $\underline{v}^-$. By Proposition 4.7(4), $\widehat{V}^u[\underline{v}^-]$ has diameter $\leq 4p_0^u$. On the other hand, $\widehat{V}^s[\underline{v}^+]$ might have big diameter. Indeed, two elements $\widehat{y},\widehat{z}\in\widehat{V}^s[\underline{v}^+]$ can have y_{-1},z_{-1} generated by different inverse branches of f and then, since the distance in $\widehat{M}$ is a supremum over all negative positions, the distance $d(\widehat{y},\widehat{z})$ can be large. This is somewhat contradictory to calling $\widehat{V}^s[\underline{v}^+]$ a stable set, but we will nevertheless use this terminology. In any case, the coding we will construct will consider intersections of stable and unstable sets, which in particular will have diameter $\leq 4p_0^u$. The next proposition justifies our terminology.

LEMMA 4.8. *The following holds for all $\varepsilon>0$ small enough.*

(1) *If $\underline{v}^+ = \{\Psi_{\widehat{x}_n}^{p_n^s,p_n^u}\}_{n\geq 0}$ is a positive ε-gpo and $d=d_s(\widehat{x}_0)$ then*

$$\widehat{V}^s[\underline{v}^+] = \vartheta^{-1}(V^s[\underline{v}^+]) = \left\{(y_n)_{n\in\mathbb{Z}}\in\widehat{M} : \begin{array}{c} y_0\in\Psi_{\widehat{x}_0}(B^d[p_0^s]\times B^{m-d}[p_0^s]) \text{ and} \\ y_n\in\Psi_{\widehat{x}_n}(B[20Q(\widehat{x}_n)]),\forall n\geq 0\end{array}\right\}.$$

(2) *If $\underline{v}^- = \{\Psi_{\widehat{x}_n}^{p_n^s,p_n^u}\}_{n\leq 0}$ is a negative ε-gpo and $d=d_s(\widehat{x}_0)$ then*

$$\widehat{V}^u[\underline{v}^-] = \left\{(y_n)_{n\in\mathbb{Z}}\in\widehat{M} : \begin{array}{c} y_0\in\Psi_{\widehat{x}_0}(B^d[p_0^u]\times B^{m-d}[p_0^u]) \text{ and} \\ y_n = f_{\widehat{x}_n}^{-1}(y_{n+1})\in\Psi_{\widehat{x}_n}(B[20Q(\widehat{x}_n)]),\forall n< 0\end{array}\right\}.$$

This lemma is just a rewriting of the definitions of $\widehat{V}^s[\underline{v}^+]$ and $\widehat{V}^u[\underline{v}^-]$ in terms of Proposition 4.7(3). We just observe that the equality $y_n = f_{\widehat{x}_n}^{-1}(y_{n+1})$ in part (2) follows from Lemma 4.2.

STABLE/UNSTABLE SETS AT v: Given an ε-double chart v, a *stable set at v* is any $\widehat{V}^s[\underline{v}^+]$ where $\underline{v}^+$ is a positive ε-gpo with $v_0=v$. We let $\widehat{\mathscr{M}}^s(v)$ denote the set of all stable sets at v. Similarly, an *unstable set at v* is any $\widehat{V}^u[\underline{v}^-]$ where $\underline{v}^-$ is a negative ε-gpo with $v_0=v$, and we let $\widehat{\mathscr{M}}^u(v)$ denote the set of all unstable sets at v.

Now we define graph transforms for stable/unstable sets. Let $v\overset{\varepsilon}{\to}w$.

GRAPH TRANSFORMS $\widehat{\mathscr{F}}^s_{v,w},\widehat{\mathscr{F}}^u_{v,w}$: The *graph transform* $\widehat{\mathscr{F}}^s_{v,w}:\widehat{\mathscr{M}}^s(w)\to\widehat{\mathscr{M}}^s(v)$ is the map that sends $\widehat{V}^s[\underline{w}^+]\in\widehat{\mathscr{M}}^s(w)$ to $\widehat{V}^s[\underline{v}^+]\in\widehat{\mathscr{M}}^s(v)$ where $\underline{v}^+=\{v_n\}_{n\geq 0}$

is defined by $v_0 = v$ and $v_n = w_{n-1}$ for all $n > 0$. Similarly, the *graph transform* $\widehat{\mathscr{F}}^u_{v,w} : \widehat{\mathscr{M}}^u(v) \to \widehat{\mathscr{M}}^u(w)$ is the map that sends $\widehat{V}^u[\underline{v}^-] \in \widehat{\mathscr{M}}^u(v)$ to $\widehat{V}^u[\underline{w}^-] \in \widehat{\mathscr{M}}^u(w)$ where $\underline{w}^- = \{w_n\}_{n\leq 0}$ is defined by $w_0 = w$ and $w_n = v_{n+1}$ for all $n < 0$.

The next result collects the main properties of the graph transforms and stable/unstable sets.

PROPOSITION 4.9. *The following holds for all $\varepsilon > 0$ small enough.*

(1) PRODUCT STRUCTURE: *If $\widehat{V}^s/\widehat{V}^u$ is a stable/unstable set at v then $\widehat{V}^s \cap \widehat{V}^u$ consists of a single element of $\widehat{M}$.*
(2) ACTION UNDER $\widehat{\mathscr{F}}^{s/u}$: *Let $v \overset{\varepsilon}{\to} w$. If $\widehat{V}^u \in \widehat{\mathscr{M}}^u(v)$ then $\widehat{\mathscr{F}}^u_{v,w}[\widehat{V}^u] \subset \widehat{f}(\widehat{V}^u)$. Similarly, if $\widehat{V}^s \in \widehat{\mathscr{M}}^s(w)$ then $\widehat{\mathscr{F}}^s_{v,w}[\widehat{V}^s] \subset \widehat{f}^{-1}(\widehat{V}^s)$.*
(3) ACTION UNDER $\widehat{f}$:

$$\widehat{f}\left(\widehat{V}^s[\{v_n\}_{n\geq 0}]\right) \subset \widehat{V}^s[\{v_n\}_{n\geq 1}] \text{ and } \widehat{f}^{-1}\left(\widehat{V}^u[\{v_n\}_{n\leq 0}]\right) \subset \widehat{V}^u[\{v_n\}_{n\leq -1}].$$

(4) HYPERBOLICITY: *If $(y_n)_{n\in\mathbb{Z}}, (z_n)_{n\in\mathbb{Z}} \in \widehat{V}^s[\underline{v}^+]$ where $\underline{v}^+ = \{\Psi^{p^s_n,p^u_n}_{\widehat{x}_n}\}_{n\geq 0}$, then*

$$d(y_n, z_n) \leq 4p^s_0 e^{-\frac{\chi}{2}n} \text{ for all } n \geq 0.$$

Hence, taking $\theta := \max\{e^{-\frac{\chi}{2}}, \frac{1}{2}\}$, if $\widehat{y}, \widehat{z} \in \widehat{V}^s[\underline{v}^+]$ then $d(\widehat{f}^n(\widehat{y}), \widehat{f}^n(\widehat{z})) < \theta^n$ for all $n \geq 0$. Similarly, if $(y_n)_{n\in\mathbb{Z}}, (z_n)_{n\in\mathbb{Z}} \in \widehat{V}^u[\underline{v}^-]$ where $\underline{v}^- = \{\Psi^{p^s_n,p^u_n}_{\widehat{x}_n}\}_{n\leq 0}$, then

$$d(y_n, z_n) \leq 4p^u_0 e^{\frac{\chi}{2}n} \text{ for all } n \leq 0.$$

Hence, if $\widehat{y}, \widehat{z} \in \widehat{V}^u[\underline{v}^-]$ then $d(\widehat{f}^n(\widehat{y}), \widehat{f}^n(\widehat{z})) < \theta^{-n}$ for all $n \leq 0$.
(5) DISJOINTNESS: *Let $v = \Psi^{p^s,p^u}_{\widehat{x}}$ and $w = \Psi^{q^s,q^u}_{\widehat{y}}$ with $\widehat{x} = \widehat{y}$. If $\widehat{V}^s, \widehat{W}^s$ are stable sets at v, w then they are either disjoint or one contains the other. The same holds for unstable sets $\widehat{V}^u, \widehat{W}^u$ at v, w.*
(6) *Let $v \overset{\varepsilon}{\to} w$. If $\widehat{V}^u$ is an unstable set at v then $\widehat{f}(\widehat{V}^u)$ intersects every stable set at w at a single element; if $\widehat{V}^s$ is a stable set at w then $\widehat{f}^{-1}(\widehat{V}^s)$ intersects every unstable set at v at a single element.*

PROOF.

(1) Write $v = \Psi^{p^s_0,p^u_0}_{\widehat{x}_0}$, $\widehat{V}^s = \widehat{V}^s[\{\Psi^{p^s_n,p^u_n}_{\widehat{x}_n}\}_{n\geq 0}]$, $\widehat{V}^u = \widehat{V}^u[\{\Psi^{p^s_n,p^u_n}_{\widehat{x}_n}\}_{n\leq 0}]$. Let us see the conditions for which $\widehat{y} = (y_n)_{n\in\mathbb{Z}}$ belongs to $\widehat{V}^s \cap \widehat{V}^u$. By the definition of stable/unstable sets, we must have that $y_0 \in V^s[\{\Psi^{p^s_n,p^u_n}_{\widehat{x}_n}\}_{n\geq 0}] \cap V^u[\{\Psi^{p^s_n,p^u_n}_{\widehat{x}_n}\}_{n\leq 0}]$. By Lemma 4.4, this latter intersection consists of a single element, hence y_0 is uniquely defined. In particular, if $n \geq 0$ then each $y_n = f^n(y_0)$ is uniquely defined. Finally, by Lemma 4.8(2), if $n \leq 0$ then $y_n = (f^{-1}_{\widehat{x}_n} \circ \cdots \circ f^{-1}_{\widehat{x}_{-1}})(y_0)$ is uniquely defined. These conditions uniquely characterize $\widehat{y}$.

(2) We prove the statement for $\widehat{\mathscr{F}}^u_{v,w}$ (the proof for $\widehat{\mathscr{F}}^s_{v,w}$ is simpler because there is no need to control inverse branches). Fix a negative ε-gpo $\underline{v}^- = \{v_n\}_{n\leq 0}$ with $v_0 = v$, and let $\underline{w}^- = \{w_n\}_{n\leq 0}$ defined by $w_0 = w$ and $w_n = v_{n+1}$ for $n < 0$. Let $V^u = V^u[\underline{v}^-]$, $\widehat{V}^u = \widehat{V}^u[\underline{v}^-]$, $W^u = V^u[\underline{w}^-]$ and $\widehat{W}^u = \widehat{V}^u[\underline{w}^-]$. By definition, $\widehat{\mathscr{F}}^u_{v,w}(\widehat{V}^u) = \widehat{W}^u$ and so we need to show that $\widehat{f}(\widehat{V}^u) \supset \widehat{W}^u$. By Lemma 4.6, we

have $\mathscr{F}^u_{v,w}(V^u) = W^u$. Since $\mathscr{F}^u_{v,w}$ is a restriction of f to a subset, $f(V^u) \supset W^u$. Writing $w_n = \Psi_{\widehat{x}_n}^{p^s_n,p^u_n}$, the definition of unstable set gives that

$$\widehat{V}^u = \left\{\widehat{y} = (y_n)_{n\in\mathbb{Z}} \in \widehat{M} : y_0 \in V^u \text{ and } y_n = f^{-1}_{\widehat{x}_{n-1}}(y_{n+1}) \text{ for all } n < 0\right\}$$

and so

$$\begin{aligned}\widehat{f}(\widehat{V}^u) &= \left\{\widehat{z} = (z_n)_{n\in\mathbb{Z}} \in \widehat{M} : z_0 \in f(V^u) \text{ and } z_n = f^{-1}_{\widehat{x}_n}(z_{n+1}) \text{ for all } n < 0\right\}\\ &\supset \left\{\widehat{z} = (z_n)_{n\in\mathbb{Z}} \in \widehat{M} : z_0 \in W^u \text{ and } z_n = f^{-1}_{\widehat{x}_n}(z_{n+1}) \text{ for all } n < 0\right\}\\ &= \widehat{W}^u.\end{aligned}$$

(3) By the definition of stable sets and Proposition 4.7(2),

$$\begin{aligned}\widehat{f}\left(\widehat{V}^s[\{v_n\}_{n\geq 0}]\right) &= \widehat{f}\left(\left\{\widehat{y} \in \widehat{M} : y_0 \in V^s[\{v_n\}_{n\geq 0}]\right\}\right)\\ &= \left\{\widehat{y} \in \widehat{M} : y_0 \in f(V^s[\{v_n\}_{n\geq 0}])\right\}\\ &\subset \left\{\widehat{y} \in \widehat{M} : y_0 \in V^s[\{v_n\}_{n\geq 1}]\right\}\\ &= \widehat{V}^s[\{v_n\}_{n\geq 1}].\end{aligned}$$

Similarly,

$$\begin{aligned}&\widehat{f}^{-1}\left(\widehat{V}^u[\{v_n\}_{n\leq 0}]\right)\\ &= \widehat{f}^{-1}\left(\left\{\widehat{y} \in \widehat{M} : y_0 \in V^u[\{v_n\}_{n\leq 0}] \text{ and } y_n = f^{-1}_{\widehat{x}_n}(y_{n+1}), \forall n < 0\right\}\right)\\ &= \left\{\widehat{y} \in \widehat{M} : y_0 \in f^{-1}_{\widehat{x}_{-1}}(V^u[\{v_n\}_{n\leq 0}]) \text{ and } y_n = f^{-1}_{\widehat{x}_{n-1}}(y_{n+1}), \forall n < 0\right\}\\ &\subset \left\{\widehat{y} \in \widehat{M} : y_0 \in V^u[\{v_n\}_{n\leq -1}] \text{ and } y_n = f^{-1}_{\widehat{x}_{n-1}}(y_{n+1}), \forall n < 0\right\}\\ &= \widehat{V}^u[\{v_n\}_{n\leq -1}].\end{aligned}$$

(4) The estimates of $d(y_n, z_n)$ follow directly from Proposition 4.7(4), so we focus on the estimates of $d(\widehat{f}^n(\widehat{y}), \widehat{f}^n(\widehat{z}))$. We let $\theta := \max\left\{e^{-\frac{\chi}{2}}, \frac{1}{2}\right\}$. Start with $\widehat{y}, \widehat{z} \in \widehat{V}^s[\underline{v}^+]$, and fix $n \geq 0$. Recalling that the diameter of M is smaller than one and using that $\widehat{f}^n(\widehat{y}) = \{y_{n+k}\}_{k\in\mathbb{Z}}$ and $\widehat{f}^n(\widehat{z}) = \{z_{n+k}\}_{k\in\mathbb{Z}}$, we have:

- If $-n \leq k \leq 0$ then $d(y_{n+k}, z_{n+k}) \leq 4p^s_0 e^{-\frac{\chi}{2}(n+k)} < \theta^{n+k}$. This implies that $2^k d(y_{n+k}, z_{n+k}) < \theta^{-k}\theta^{n+k} = \theta^n$.
- If $k < -n$ then $2^k d(y_{n+k}, z_{n+k}) < 2^k \leq \theta^{-k} < \theta^n$.

Hence $d(\widehat{f}^n(\widehat{y}), \widehat{f}^n(\widehat{z})) < \theta^n$. The other estimate is simpler. Let $\widehat{y}, \widehat{z} \in \widehat{V}^u[\underline{v}^-]$ and fix $n < 0$. For $k \leq 0$ we have $d(y_{n+k}, z_{n+k}) \leq 4p^u_0 e^{\frac{\chi}{2}(n+k)} < \theta^{-n-k}$ and so $2^k d(y_{n+k}, z_{n+k}) < \theta^{-k}\theta^{-n-k} = \theta^{-n-2k} \leq \theta^{-n}$. Hence $d(\widehat{f}^n(\widehat{y}), \widehat{f}^n(\widehat{z})) < \theta^{-n}$.

(5) The proof is motivated by [**Sar13**, Proposition 6.4], with the modifications needed to deal with inverse branches. We start dealing with unstable sets. Let $\widehat{V}^u = \widehat{V}^u[\underline{v}^-]$ where $\underline{v}^- = \{\Psi_{\widehat{x}_n}^{p^s_n,p^u_n}\}_{n\leq 0}$ and $\widehat{W}^u = \widehat{V}^u[\underline{w}^-]$ where $\underline{w}^- = \{\Psi_{\widehat{y}_n}^{q^s_n,q^u_n}\}_{n\leq 0}$, and assume that $\widehat{x}_0 = \widehat{y}_0$ and $\widehat{V}^u \cap \widehat{W}^u \neq \emptyset$. We will show that if $q^u_0 \leq p^u_0$ then $\widehat{W}^u \subset \widehat{V}^u$. To do this, we first prove that for $n < 0$ the inverse branches $f^{-1}_{\widehat{x}_n}$ and $f^{-1}_{\widehat{y}_n}$ coincide in the domains of interest.

CLAIM 1: For every $n \leq 0$ it holds $\vartheta_n[\widehat{W}^u] \subset E_{\widehat{x}_{n-1}}$ and $\vartheta_{n-1}[\widehat{W}^u] \subset f_{\widehat{x}_{n-1}}^{-1}(E_{\widehat{x}_{n-1}})$.

PROOF OF CLAIM 1. Fix $\widehat{z} = (z_k)_{k\in\mathbb{Z}} \in \widehat{V}^u \cap \widehat{W}^u$ and $n \leq 0$. Recall, by the proof of Lemma 4.2, that $Q(\widehat{x}_n) < \varepsilon\mathfrak{r}(\widehat{x}_{n-1})$. Using that $\widehat{z} \in \widehat{V}^u$, Lemma 4.8 implies that

$$z_n \in B(\vartheta_0[\widehat{x}_n], 40Q(\widehat{x}_n)) \subset B(\vartheta_1[\widehat{x}_{n-1}], 50Q(\widehat{x}_n)),$$

since by the overlap assumption we have $d(\vartheta_0[\widehat{x}_n], \vartheta_1[\widehat{x}_{n-1}]) \ll p_n^s \wedge p_n^u < Q(\widehat{x}_n)$. Now let $\widehat{t} = (t_k)_{k\in\mathbb{Z}} \in \widehat{W}^u$ be arbitrary. Using part (4) and the inequality $p_0^u \leq e^{-\varepsilon n}p_n^u$, if $\varepsilon > 0$ is small then $d(z_n, t_n) \leq 4q_0^u e^{\frac{\chi}{2}n} \leq 4q_0^u e^{\varepsilon n} \leq 4\left(\frac{q_0^u}{p_0^u}\right)p_n^u \leq 4p_n^u$ and so $\vartheta_n[\widehat{W}^u] \subset B(\vartheta_1[\widehat{x}_{n-1}], 50Q(\widehat{x}_n) + 4p_n^u) \subset E_{\widehat{x}_{n-1}}$, where in the last passage we used that $50Q(\widehat{x}_n) + 4p_n^u < 60Q(\widehat{x}_n) < 60\varepsilon\mathfrak{r}(\widehat{x}_{n-1}) < 2\mathfrak{r}(\widehat{x}_{n-1})$.

The second inclusion is easier. We have $z_{n-1} \in B(\vartheta_0[\widehat{x}_{n-1}], 40Q(\widehat{x}_{n-1}))$ and $d(z_{n-1}, t_{n-1}) \leq 4p_{n-1}^u$, thus $\vartheta_{n-1}[\widehat{W}^u] \subset B(\vartheta_0[\widehat{x}_{n-1}], 50Q(\widehat{x}_{n-1}))$. Now, (A6) implies $f_{\widehat{x}_{n-1}}^{-1}(E_{\widehat{x}_{n-1}}) \supset B(\vartheta_0[\widehat{x}_{n-1}], 2d(\vartheta_0[\widehat{x}_{n-1}], \mathscr{S})^a\mathfrak{r}(\widehat{x}_{n-1}))$. The radius of this latter ball is $> 2\rho(\widehat{x}_{n-1})^{2a} > 50Q(\widehat{x}_{n-1})$, therefore $\vartheta_{n-1}[\widehat{W}^u] \subset f_{\widehat{x}_{n-1}}^{-1}(E_{\widehat{x}_{n-1}})$. □

Letting $d = d_s(\widehat{x}_0)$, Claim 1 implies that

$$\widehat{W}^u = \left\{ (z_n)_{n\in\mathbb{Z}} \in \widehat{M} : \begin{array}{c} z_0 \in \Psi_{\widehat{x}_0}(B^d[q_0^u] \times B^{m-d}[q_0^u]) \text{ and} \\ z_n = f_{\widehat{x}_n}^{-1}(z_{n+1}) \in \Psi_{\widehat{y}_n}(B[20Q(\widehat{y}_n)]), \forall n < 0 \end{array} \right\},$$

i.e. we can substitute $f_{\widehat{y}_n}^{-1}$ by $f_{\widehat{x}_n}^{-1}$ in the inverse branches. This allows us to employ the same methods of [**Sar13**, Proposition 6.4], summarized in our context as follows. Below, "n is small enough" means that $-n$ is large enough.

CLAIM 2: If n is small enough then $\vartheta_n[\widehat{V}^u] \subset \Psi_{\widehat{x}_n}\left(B^d[\tfrac{1}{2}p_n^u] \times B^{m-d}[\tfrac{1}{2}p_n^u]\right)$.

The proof is made exactly as in [**Sar13**, Claim 1, p. 420].

CLAIM 3: If n is small enough then $\vartheta_n[\widehat{W}^u] \subset \Psi_{\widehat{x}_n}\left(B^d[p_n^u] \times B^{m-d}[p_n^u]\right)$.

The proof of this claim is made exactly as in [**Sar13**, Claim 2, p. 420]. Finally, to prove that $\widehat{W}^u \subset \widehat{V}^u$, we proceed as in [**BO18**, Claim 3, p. 102]. The proof for stable sets is simpler and can be made as in [**Sar13**, Proposition 6.4] or [**BO18**, Claim 3, p. 102], since in this case we do not have difficulties with inverse branches.

(6) Since $\widehat{f}$ is invertible, it is enough to prove the first statement. Let $v = \Psi_{\widehat{x}}^{p^s,p^u}$, $w = \Psi_{\widehat{y}}^{q^s,q^u}$, $\widehat{V}^u = \widehat{V}^u[\underline{v}^-]$ an unstable set at v, and $\widehat{V}^s = \widehat{V}^s[\underline{w}^+]$ a stable set at w. Let $V^u = V^u[\underline{v}^-]$ and $V^s = V^s[\underline{w}^+]$. If $(z_n)_{n\in\mathbb{Z}} \in \widehat{f}(\widehat{V}^u) \cap \widehat{V}^s$ then $z_0 \in f(V^u) \cap V^s$. By Proposition 4.5(3), z_0 is uniquely defined. In particular, z_n is uniquely defined for all $n \geq 0$. Furthermore, $z_{-1} = f_{\widehat{x}}^{-1}(z_0)$ is uniquely defined and therefore, by Lemma 4.8(2), z_n is uniquely defined for each $n < -1$. This show that $\widehat{f}(\widehat{V}^u) \cap \widehat{V}^s$ consists of a single element. □

SHADOWING: We say that an ε-gpo $\{\Psi_{\widehat{x}_n}^{p_n^s,p_n^u}\}_{n\in\mathbb{Z}}$ *shadows* a point $\widehat{y} = (y_n)_{n\in\mathbb{Z}} \in \widehat{M}$ if $y_n \in \Psi_{\widehat{x}_n}(B[20Q(\widehat{x}_n)])$ for all $n \in \mathbb{Z}$.

LEMMA 4.10. *Every ε-gpo shadows a unique point.*

The proof is the same as [**LS19**, Theorem 4.2], with the shadowed point of an ε-gpo $\underline{v}$ being the intersection $\widehat{V}^s[\underline{v}] \cap \widehat{V}^u[\underline{v}]$. In particular, by Lemma 4.4 we obtain that shadowing implies the stronger inclusion $y_n \in \Psi_{\widehat{x}_n}(B[p_n^s \wedge p_n^u])$ for all $n \in \mathbb{Z}$.

4.7. Hyperbolicity along stable/unstable sets

In this section, we establish some estimates that will play an essential role in the proof of Theorem 6.1. Given a positive ε-gpo $\underline{v}^+$, let $V^s/\widehat{V}^s$ be its s-admissible manifold/set. Below, we use the notation $T\widehat{V}^s = \widehat{TV^s}$ to represent the tangent bundle of V^s as a subset of $\widehat{TM}$, i.e. for every $\widehat{y} \in \widehat{V}^s$ we let $T_{\widehat{y}}\widehat{V}^s$ be the tangent subspace $T_{\vartheta[\widehat{y}]}V^s$ as a subset of $\widehat{TM}_{\widehat{y}}$.

PROPOSITION 4.11. *The following holds for $\varepsilon > 0$ small enough.*

(1) *If $\widehat{V}^s = \widehat{V}^s[\underline{v}^+]$ where $\underline{v}^+ = \{\Psi_{\widehat{x}_n}^{p_n^s,p_n^u}\}_{n\geq 0}$ is a positive ε-gpo, then for every $v \in T\widehat{V}^s$ with $\|v\| = 1$ and every $n \geq 0$ it holds*

$$\|\widehat{df}^{(n)}v\| \leq 8\|C(\widehat{x}_0)^{-1}\|e^{-\frac{\chi}{2}n}.$$

Similarly, if $\widehat{V}^u = \widehat{V}^u[\underline{v}^-]$ where $\underline{v}^- = \{\Psi_{\widehat{x}_n}^{p_n^s,p_n^u}\}_{n\leq 0}$ is a negative ε-gpo, then for every $v \in T\widehat{V}^u$ with $\|v\| = 1$ and every $n \geq 0$ it holds

$$\|\widehat{df}^{(-n)}v\| \leq 8\|C(\widehat{x}_0)^{-1}\|e^{-\frac{\chi}{2}n}.$$

(2) *If $\underline{v}$ is an ε-gpo and $\{\widehat{y}\} = \widehat{V}^s[\underline{v}] \cap \widehat{V}^u[\underline{v}]$, then for all nonzero $v \in T_{\widehat{y}}\widehat{V}^s$ it holds*

$$\liminf_{n\to+\infty} \tfrac{1}{n} \log \|\widehat{df}^{(-n)}v\| > 0$$

and for all nonzero $v \in T_{\widehat{y}}\widehat{V}^u$ it holds

$$\liminf_{n\to+\infty} \tfrac{1}{n} \log \|\widehat{df}^{(n)}v\| > 0.$$

PROOF. All statements are about the cocycle $(\widehat{df}^{(n)})_{n\in\mathbb{Z}}$, therefore we can prove the result in the domains of the Pesin charts and then push it to the respective tangent spaces. As we will see, the proof is symmetric and so we just need to prove the estimates for the stable manifolds.

(1) Let $d = d_s(\widehat{x}_0)$, and let G_k be the representing function of $V^s[\{\Psi_{\widehat{x}_n}^{p_n^s,p_n^u}\}_{n\geq k}]$. Fix $v \in T_{\widehat{y}}\widehat{V}^s$ with $\|v\| = 1$, and let $z = (v_0, G_0(v_0))$ with $\vartheta[\widehat{y}] = y = \Psi_{\widehat{x}_0}(z)$. Thus

$$v = (d\Psi_{\widehat{x}_0})_z \begin{bmatrix} w \\ (dG_0)_{v_0}w \end{bmatrix}$$

for some $w \in \mathbb{R}^d$. In particular, $\|w\| \leq \left\|\begin{bmatrix} w \\ (dG_0)_{v_0}w \end{bmatrix}\right\| \leq 2\|C(\widehat{x}_0)^{-1}\|$. Using that $\Psi_{\widehat{x}_n} \circ F_{\widehat{x}_{n-1},\widehat{x}_n} \circ \cdots \circ F_{\widehat{x}_0,\widehat{x}_1} = f^n \circ \Psi_{\widehat{x}_0}$, we have

$$\widehat{df}^{(n)}v = [d(\Psi_{\widehat{x}_n} \circ F_{\widehat{x}_{n-1},\widehat{x}_n} \circ \cdots \circ F_{\widehat{x}_0,\widehat{x}_1})]_z \begin{bmatrix} w \\ (dG_0)_{v_0}w \end{bmatrix}$$

and so

$$\|\widehat{df}^{(n)}v\| \leq 2\left\|[d(F_{\widehat{x}_{n-1},\widehat{x}_n} \circ \cdots \circ F_{\widehat{x}_0,\widehat{x}_1})]_z \begin{bmatrix} w \\ (dG_0)_{v_0}w \end{bmatrix}\right\|.$$

Define inductively $z_k = (v_k, G_k(v_k))$ and $w_k \in \mathbb{R}^d$ by $z_0 = z$, $z_k = F_{\widehat{x}_{k-1},\widehat{x}_k}(z_{k-1})$, $w_0 = w$ and

$$\begin{bmatrix} w_k \\ (dG_k)_{v_k}w_k \end{bmatrix} = (dF_{\widehat{x}_{k-1},\widehat{x}_k})_{z_{k-1}} \begin{bmatrix} w_{k-1} \\ (dG_{k-1})_{v_{k-1}}w_{k-1} \end{bmatrix}.$$

In particular, $z_k \in B^d[p_k^s] \times B^{m-d}[p_k^s] \subset B[2p_k^s]$. We want to bound

$$\left\| \begin{bmatrix} w_n \\ (dG_n)_{v_n} w_n \end{bmatrix} \right\|.$$

By Lemma A.2(1), we have

$$\left\| \begin{bmatrix} w_k \\ (dG_k)_{v_k} w_k \end{bmatrix} \right\| \leq 2\|w_k\|,$$

hence it is enough to estimate $\|w_n\|$. Write $F_{\widehat{x}_{k-1},\widehat{x}_k} = D(\widehat{x}_{k-1}) + H_{k-1}^+$, where H_{k-1}^+ satisfies Theorem 4.3. Then

$$\begin{aligned} \|w_k\| &\leq \|D_s(\widehat{x}_{k-1})w_{k-1}\| + \left\|(dH_{k-1}^+)_{z_{k-1}} \begin{bmatrix} w_{k-1} \\ (dG_{k-1})_{v_{k-1}} w_{k-1} \end{bmatrix}\right\| \\ &\leq \left(\|D_s(\widehat{x}_{k-1})\| + 2\left\|(dH_{k-1}^+)_{z_{k-1}}\right\|\right)\|w_{k-1}\| \end{aligned}$$

and so, by Lemma 3.1(2) and Theorem 4.3(b), we obtain that

$$\|w_k\| \leq \left[e^{-\chi} + 4\varepsilon(p_{k-1}^s)^{\beta/3}\right]\|w_{k-1}\| \leq e^{-\frac{\chi}{2}}\|w_{k-1}\|, \tag{4.2}$$

where in the last passage we used that $\varepsilon > 0$ is small enough. Therefore, we conclude that $\|w_n\| \leq e^{-\frac{\chi}{2}n}\|w_0\| \leq 2\left\|C(\widehat{x}_0)^{-1}\right\|e^{-\frac{\chi}{2}n}$ and so $\|\widehat{df}^{(n)}v\| \leq 8\left\|C(\widehat{x}_0)^{-1}\right\|e^{-\frac{\chi}{2}n}$.

(2) Write $\underline{v} = \{v_n\}_{n\in\mathbb{Z}} = \{\Psi_{\widehat{x}_n}^{p_n^s,p_n^u}\}_{n\in\mathbb{Z}}$ with $\eta_n = p_n^s \wedge p_n^u$, and let $\widehat{y} \in \widehat{V}^s[\underline{v}] \cap \widehat{V}^u[\underline{v}]$. Letting $\widehat{V}_n^s = \widehat{V}^s[\{v_k\}_{k\geq n}]$ and $\widehat{V}_n^u = \widehat{V}^u[\{v_k\}_{k\leq n}]$, we have $\{\widehat{f}^n(\widehat{y})\} = \widehat{V}_n^s \cap \widehat{V}_n^u$. Fix a nonzero vector $v \in T_{\widehat{y}}\widehat{V}^s$ and $n \geq 0$. Then $w = \widehat{df}^{(-n)}v \in T_{\widehat{f}^{-n}(\widehat{y})}\widehat{V}_{-n}^s$. By part (1), we have $\|v\| = \|\widehat{df}^{(n)}w\| \leq 8\left\|C(\widehat{x}_{-n})^{-1}\right\|e^{-\frac{\chi}{2}n}\|w\|$ and so $\|\widehat{df}^{(-n)}v\| \geq \frac{1}{8}\|v\|e^{\frac{\chi}{2}n}\left\|C(\widehat{x}_{-n})^{-1}\right\|^{-1}$. Since $\left\|C(\widehat{x}_{-n})^{-1}\right\|^{-1} \geq \eta_{-n}^{\beta/48} \geq (e^{-\varepsilon n}\eta_0)^{\beta/48}$, we get that

$$\|\widehat{df}^{(-n)}v\| \geq \tfrac{1}{8}\eta_0^{\beta/48}\|v\|e^{(\frac{\chi}{2}-\frac{\beta\varepsilon}{48})n}$$

and so $\liminf\limits_{n\to+\infty} \frac{1}{n}\log\|\widehat{df}^{(-n)}v\| \geq \frac{\chi}{2} - \frac{\beta\varepsilon}{48} > 0$ if $\varepsilon > 0$ is small enough.

□

Observe that part (1) can be improved if we have an estimate that is better than the one provided by Theorem 4.3(b), which in general is not true for ε-gpo's. Nevertheless, if we apply the method to a real orbit, then we can apply Theorem 3.3 and, the smaller $\|z_{k-1}\|$ is, the closer we get to a contraction of rate $e^{-\chi}$. This is the content of the next result, which is the claim in [**BO18**, p. 85], adapted to our context. Fix $\widehat{x} \in \text{NUH}^*$, and let $\widehat{x}_n = \widehat{f}^n(\widehat{x})$, $p_n^s = q^s(\widehat{x}_n)$ and $p_n^u = q^u(\widehat{x}_n)$. We know that $\{\Psi_{\widehat{x}_n}^{p_n^s,p_n^u}\}_{n\in\mathbb{Z}}$ is an ε-gpo. Fix $\chi' \in (0,\chi)$, let $\delta = (e^{-\chi'} - e^{-\chi})^{3/\beta} < 1$ and consider the sequence $\underline{v} = \{\Psi_{\widehat{x}_n}^{\delta p_n^s,\delta p_n^u}\}_{n\in\mathbb{Z}}$. In practice, δ diminishes the domains of the Pesin charts. Observe that $\underline{v}$ is *not necessarily* an ε-gpo, since the parameters $\delta p_n^s, \delta p_n^u$ might not satisfy (GPO2). Nevertheless, they do satisfy condition (GT1) in the Appendix, so we can define graph transforms. Hence, although $\underline{v}$ is not an ε-gpo, this sequence still defines invariant sets, which we will denote by $\widehat{V}_{\chi'}^s = \widehat{V}_{\chi'}^s[\underline{v}^+]$ and $\widehat{V}_{\chi'}^u = \widehat{V}_{\chi'}^u[\underline{v}^-]$. Now, since we diminished the domains of the Pesin charts then the estimates used in the proof of Proposition 4.11(1) can be improved to give a rate of contraction of the order of $e^{-\chi'}$.

COROLLARY 4.12. *Let* $\widehat{x} \in \text{NUH}^*$. *For a fixed* $\chi' \in (0, \chi)$, *let* $\underline{v}$ *be as above. If* $\widehat{V}^s_{\chi'} = \widehat{V}^s_{\chi'}[\underline{v}^+]$ *then for every* $v \in T\widehat{V}^s_{\chi'}$ *with* $\|v\| = 1$ *and every* $n \geq 0$ *it holds*

$$\|\widehat{df}^{(n)} v\| \leq 8\|C(\widehat{x})^{-1}\|e^{-\chi' n}.$$

Similarly, if $\widehat{V}^u_{\chi'} = \widehat{V}^u_{\chi'}[\underline{v}^-]$ *then for every* $v \in T\widehat{V}^u_{\chi'}$ *with* $\|v\| = 1$ *and every* $n \geq 0$ *it holds*

$$\|\widehat{df}^{(-n)} v\| \leq 8\|C(\widehat{x})^{-1}\|e^{-\chi' n}.$$

PROOF. Proceed exactly as in the proof of Proposition 4.11(1), except at inequalities (4.2) which, by Theorem 3.3(3)(b), take the stronger form

$$\|w_k\| \leq \left[e^{-\chi} + 4\varepsilon(\delta p^s_{k-1})^{\beta/3}\right] \|w_{k-1}\| \leq \left[e^{-\chi} + \delta^{\beta/3}\right] \|w_{k-1}\| = e^{-\chi'} \|w_{k-1}\|,$$

thus establishing the result. □

Observe that the closer χ' is to χ, the better the estimate gets but the domains where the estimates hold will shrink to a point.

CHAPTER 5

Coarse graining

We now construct a countable family of ε-double charts that defines a topological Markov shift that shadows all relevant orbits. Recall the definitions of the sets NUH, NUH*, and NUH$^\#$.

THEOREM 5.1. *For all $\varepsilon > 0$ sufficiently small, there exists a countable family $\mathscr{A}$ of ε-double charts with the following properties:*

(1) DISCRETENESS: *For all $t > 0$, the set $\{\Psi_{\widehat{x}}^{p^s,p^u} \in \mathscr{A} : p^s, p^u > t\}$ is finite.*
(2) SUFFICIENCY: *If $\widehat{x} \in \mathrm{NUH}^\#$ then there is a recurrent sequence $\underline{v} \in \mathscr{A}^{\mathbb{Z}}$ that shadows x.*
(3) RELEVANCE: *For all $v \in \mathscr{A}$ there is an ε-gpo $\underline{v} \in \mathscr{A}^{\mathbb{Z}}$ with $v_0 = v$ that shadows a point in* NUH$^\#$.

Above, $(v_n)_{n\in\mathbb{Z}} \in \mathscr{A}^{\mathbb{Z}}$ is called *recurrent* if there are $v, w \in \mathscr{A}$ s.t. $v_n = v$ for infinitely many $n > 0$ and $v_n = w$ for infinitely many $n < 0$. Parts (1) and (3) will be crucial to prove the inverse theorem (Theorem 6.1). Since NUH$^\#$ carries all f-adapted χ-hyperbolic measures (see Lemma 3.6), part (2) says that the ε-gpo's in $\mathscr{A}$ shadow a.e. point with respect to every f-adapted χ-hyperbolic measure.

PROOF. We follow the same strategy of [**LM18**, Theorem 5.1]. For $t > 0$, let $M_t = \{x \in M : d(x, \mathscr{S}) \geq t\}$. Since M has finite diameter (remember we are even assuming it is smaller than one), each M_t is precompact. Let $\mathbb{N}_0 = \mathbb{N} \cup \{0\}$. Fix a countable open cover $\mathscr{P} = \{\mathfrak{B}_i\}_{i\in\mathbb{N}_0}$ of $M\backslash\mathscr{S}$ s.t.:

- $\mathfrak{B}_i := \mathfrak{B}_{z_i} = B(z_i, 2\mathfrak{d}(z_i))$ for some $z_i \in M$.
- For every $t > 0$, $\{\mathfrak{B} \in \mathscr{P} : \mathfrak{B} \cap M_t \neq \emptyset\}$ is finite.

Let $X := M^3 \times \mathrm{GL}(m,\mathbb{R})^3 \times (0,1]$. For $\widehat{x} \in \mathrm{NUH}^\#$, let $\Gamma(\widehat{x}) = (\underline{\widehat{x}}, \underline{C}, \underline{Q}) \in X$ with

$$\underline{\widehat{x}} = (\vartheta_{-1}[\widehat{x}], \vartheta_0[\widehat{x}], \vartheta_1[\widehat{x}]),\ \underline{C} = (C(\widehat{f}^{-1}(\widehat{x})), C(\widehat{x}), C(\widehat{f}(\widehat{x}))),\ \underline{Q} = Q(\widehat{x}).$$

Let $Y = \{\Gamma(\widehat{x}) : \widehat{x} \in \mathrm{NUH}_\chi^\#\}$. We want to construct a countable dense subset of Y. Since the maps $\widehat{x} \mapsto C(\widehat{x}), Q(\widehat{x})$ are usually just measurable, we apply a precompactness argument. For each triple of vectors $\underline{k} = (k_{-1}, k_0, k_1)$, $\underline{\ell} = (\ell_{-1}, \ell_0, \ell_1)$, $\underline{a} = (a_{-1}, a_0, a_1) \subset \mathbb{N}_0^3$, $b \in \mathbb{N}_0$ and $d \in \{0, 1, \ldots, m\}$, define

$$Y_{\underline{k},\underline{\ell},\underline{a},b,d} := \left\{\Gamma(\widehat{x}) \in Y : \begin{array}{cc} e^{-k_i-1} \leq d(\vartheta_i[\widehat{x}], \mathscr{S}) < e^{-k_i}, & -1 \leq i \leq 1 \\ e^{\ell_i} \leq \|C(\widehat{f}^i(\widehat{x}))^{-1}\| < e^{\ell_i+1}, & -1 \leq i \leq 1 \\ \vartheta_i[\widehat{x}] \in \mathfrak{B}_{a_i}, & -1 \leq i \leq 1 \\ e^{-b-1} \leq Q(\widehat{x}) < e^{-b} & \\ d_s(\widehat{x}) = d & \end{array}\right\}.$$

Clearly, any dependence on $d \in \{0, 1, \ldots, m\}$ is finite. To avoid an extra summand in the calculations, in the sequel we will omit the dependence on d.

CLAIM 1: $Y=\bigcup_{\substack{\underline{k},\underline{\ell},\underline{a}\in\mathbb{N}_0^3\\ b,d\in\mathbb{N}_0}} Y_{\underline{k},\underline{\ell},\underline{a},b,d}$, and each $Y_{\underline{k},\underline{\ell},\underline{a},b,d}$ is precompact in X.

Proof of Claim 1. The first statement is clear. We focus on the second. Fix $\underline{k},\underline{\ell},\underline{a}\in\mathbb{N}_0^3$, $b,d\in\mathbb{N}_0$. Take $\Gamma(\widehat{\underline{x}})\in Y_{\underline{k},\underline{\ell},\underline{a},b,d}$. Then

$$\widehat{\underline{x}}\in M_{e^{-k_{-1}-1}}\times M_{e^{-k_0-1}}\times M_{e^{-k_1-1}},$$

a precompact subset of M^3. For $|i|\leq 1$, $C(\widehat{f^i}(\widehat{\underline{x}}))$ is an element of $\mathrm{GL}(m,\mathbb{R})$ with norm ≤ 1 and inverse norm $\leq e^{\ell_i+1}$, hence it belongs to a compact subset of $\mathrm{GL}(m,\mathbb{R})$. This guarantees that $\underline{C}$ belongs to a compact subset of $\mathrm{GL}(m,\mathbb{R})^3$. Also, $\underline{Q}\in[e^{-b-1},1]$, a compact subinterval of $(0,1]$. Since the product of precompact sets is precompact, the claim is proved.

Let $j\geq 0$. By Claim 1, there exists a finite set $Y_{\underline{k},\underline{\ell},\underline{a},b,d}(j)\subset Y_{\underline{k},\underline{\ell},\underline{a},b,d}$ s.t. for every $\Gamma(\widehat{\underline{x}})\in Y_{\underline{k},\underline{\ell},\underline{a},b,d}$ there exists $\Gamma(\widehat{\underline{y}})\in Y_{\underline{k},\underline{\ell},\underline{a},b,d}(j)$ s.t.:

(a) $d(\vartheta_i[\underline{x}],\vartheta_i[\underline{y}])+\left\|\widetilde{C(\widehat{f^i}(\widehat{\underline{x}}))}-\widetilde{C(\widehat{f^i}(\widehat{\underline{y}}))}\right\|<e^{-8(j+2)}$ for $-1\leq i\leq 1$.

(b) $\frac{Q(\widehat{\underline{x}})}{Q(\widehat{\underline{y}})}=e^{\pm\varepsilon/3}$.

THE ALPHABET $\mathscr{A}$: Let $\mathscr{A}$ be the countable family of $\Psi_{\widehat{\underline{x}}}^{p^s,p^u}$ s.t.:

(CG1) $\Gamma(\widehat{\underline{x}})\in Y_{\underline{k},\underline{\ell},\underline{a},b,d}(j)$ for some $(\underline{k},\underline{\ell},\underline{a},b,d,j)\in\mathbb{N}_0^3\times\mathbb{N}_0^3\times\mathbb{N}_0^3\times\mathbb{N}_0\times\mathbb{N}_0\times\mathbb{N}_0$.
(CG2) $0<p^s,p^u\leq\delta_\varepsilon Q(\widehat{\underline{x}})$ and $p^s,p^u\in I_\varepsilon$.
(CG3) $e^{-j-2}\leq p^s\wedge p^u\leq e^{-j+2}$.

Proof of discreteness. By Lemma 3.1(4), we have:

$$\|C(\widehat{f^{\pm1}}(\widehat{\underline{x}}))^{-1}\|\leq\rho(\widehat{\underline{x}})^{-2a}(1+e^{2\chi})^{1/2}\|C(\widehat{\underline{x}})^{-1}\|. \tag{5.1}$$

Fix $t>0$, and let $\Psi_{\widehat{\underline{x}}}^{p^s,p^u}\in\mathscr{A}$ with $p^s,p^u>t$. Note that $\rho(\widehat{\underline{x}})>\rho(\widehat{\underline{x}})^{2a}>Q(\widehat{\underline{x}})>p^s,p^u>t$. If $\Gamma(\widehat{\underline{x}})\in Y_{\underline{k},\underline{\ell},\underline{a},b,d}(j)$ then:

- Finiteness of $\underline{k}$: for $|i|\leq 1$, $e^{-k_i}>d(\vartheta_i[\underline{x}],\mathscr{S})\geq\rho(\widehat{\underline{x}})>t$, hence $k_i<|\log t|$.
- Finiteness of $\underline{\ell}$: we have $e^{\ell_0}\leq\|C(\widehat{\underline{x}})^{-1}\|<Q(\widehat{\underline{x}})^{-1}<t^{-1}$, hence $\ell_0<|\log t|$. For $i=\pm1$, inequality (5.1) implies that
$$e^{\ell_i}\leq\|C(\widehat{f^i}(\widehat{\underline{x}}))^{-1}\|<t^{-1}(1+e^{2\chi})^{1/2}t^{-1}<e^{\chi+1}t^{-2},$$
hence $\ell_i<\chi+1+2|\log t|=:T_t$, which is bigger than $|\log t|$.
- Finiteness of b: $e^{-b}>Q(\widehat{\underline{x}})>t$, hence $b<|\log t|$.
- Finiteness of j: $t<p^s\wedge p^u\leq e^{-j+2}$, hence $j\leq|\log t|+2$.
- Finiteness of $\underline{a}$: $\vartheta_i[\underline{x}]\in\mathfrak{B}_{a_i}\cap M_t$, hence $\mathfrak{B}_{a_i}$ belongs to the finite set $\{\mathfrak{B}\in\mathscr{P}:\mathfrak{B}\cap M_t\neq\emptyset\}$.
- Finiteness of (p^s,p^u): $t<p^s,p^u$, hence $\#\{(p^s,p^u):p^s,p^u>t\}\leq\#(I_\varepsilon\cap(t,1])^2$ is finite.

The first four items above give that, for $\underline{a}\in\mathbb{N}_0^3$ and $t>0$,

$$\#\left\{\Gamma(\widehat{\underline{x}}):\begin{array}{c}\Psi_{\widehat{\underline{x}}}^{p^s,p^u}\in\mathscr{A}\text{ s.t. }p^s,p^u>t\\ \text{and }\vartheta_i[\underline{x}]\in\mathfrak{B}_{a_i},|i|\leq 1\end{array}\right\}\leq\sum_{j=0}^{\lceil|\log t|\rceil+2}\sum_{b=0}^{\lceil|\log t|\rceil}\sum_{\substack{k_i,\ell_i=0\\-1\leq i\leq 1}}^{T_t}\#Y_{\underline{k},\underline{\ell},\underline{a},b,d}(j)$$

is the finite sum of finite terms, hence finite. Together with the last two items, we conclude that

$$\#\left\{\Psi_{\widehat{x}}^{p^s,p^u} \in \mathscr{A} : p^s, p^u > t\right\} \leq \sum_{j=0}^{\lceil|\log t|\rceil+2} \sum_{b=0}^{\lceil|\log t|\rceil} \sum_{\substack{k_i,\ell_i=0 \\ -1\leq i\leq 1}}^{T_t} \#Y_{\underline{k},\underline{\ell},\underline{a},b,d}(j)$$
$$\times (\#\{\mathfrak{B} \in \mathscr{P} : \mathfrak{B} \cap M_t \neq \emptyset\})^3 \times (\#(I_\varepsilon \cap (t,1]))^2$$

is finite. This proves the discreteness of $\mathscr{A}$.

Proof of sufficiency. Let $\widehat{x} \in \mathrm{NUH}^\#$, and take $(k_i)_{i\in\mathbb{Z}}$, $(\ell_i)_{i\in\mathbb{Z}}$, $(a_i)_{i\in\mathbb{Z}}$, $(b_i)_{i\in\mathbb{Z}}$, $(d_i)_{i\in\mathbb{Z}}$, $(j_i)_{i\in\mathbb{Z}}$ s.t.:

$$d(\vartheta_i[\widehat{x}], \mathscr{S}) \in [e^{-k_i-1}, e^{-k_i}),\ \|C(\widehat{f^i}(\widehat{x}))^{-1}\| \in [e^{\ell_i}, e^{\ell_i+1}),\ \vartheta_i[\widehat{x}] \in \mathfrak{B}_{a_i},$$
$$Q(\widehat{f^i}(\widehat{x})) \in [e^{-b_i-1}, e^{-b_i}),\ d_i = d_s(\widehat{f^i}(\widehat{x})),\ q(\widehat{f^i}(\widehat{x})) \in [e^{-j_i-1}, e^{-j_i+1}).$$

For $n \in \mathbb{Z}$, define

$$\underline{k}^{(n)} = (k_{n-1}, k_n, k_{n+1}),\ \underline{\ell}^{(n)} = (\ell_{n-1}, \ell_n, \ell_{n+1}),\ \underline{a}^{(n)} = (a_{n-1}, a_n, a_{n+1}).$$

Then $\Gamma(\widehat{f^n}(\widehat{x})) \in Y_{\underline{k}^{(n)},\underline{\ell}^{(n)},\underline{a}^{(n)},b_n,d_n}$. Take $\Gamma(\widehat{x}_n) \in Y_{\underline{k}^{(n)},\underline{\ell}^{(n)},\underline{a}^{(n)},b_n,d_n}(j_n)$ s.t.:

(a$_n$) $d(\vartheta_i[\widehat{f^n}(\widehat{x})], \vartheta_i[\widehat{x}_n]) + \left\|\widetilde{C(\widehat{f^i}(\widehat{f^n}(\widehat{x})))} - \widetilde{C(\widehat{f^i}(\widehat{x}_n))}\right\| < e^{-8(j_n+2)}$ for $|i| \leq 1$.

(b$_n$) $\frac{Q(\widehat{f^n}(\widehat{x}))}{Q(\widehat{x}_n)} = e^{\pm\varepsilon/3}$.

Define $p_n^s = \delta_\varepsilon \min\{e^{\varepsilon|k|}Q(\widehat{x}_{n+k}) : k \geq 0\}$ and $p_n^u = \delta_\varepsilon \min\{e^{\varepsilon|k|}Q(\widehat{x}_{n+k}) : k \leq 0\}$. We claim that $\{\Psi_{\widehat{x}_n}^{p_n^s,p_n^u}\}_{n\in\mathbb{Z}}$ is an ε-gpo in $\mathscr{A}^{\mathbb{Z}}$ that shadows x.

Claim 2: $\Psi_{\widehat{x}_n}^{p_n^s,p_n^u} \in \mathscr{A}$ for all $n \in \mathbb{Z}$.

(CG1) By definition, $\Gamma(\widehat{x}_n) \in Y_{\underline{k}^{(n)},\underline{\ell}^{(n)},\underline{a}^{(n)},b_n,d_n}(j_n)$.
(CG2) By (b$_n$), $\inf\{e^{\varepsilon|k|}Q(\widehat{x}_{n+k}) : k \geq 0\} = e^{\pm\varepsilon/3}\inf\{e^{\varepsilon|k|}Q(\widehat{f^{n+k}}(\widehat{x})) : k \geq 0\}$ is positive. Since the only accumulation point of I_ε is zero, it follows that p_n^s is well-defined and positive. The same proof applies to p_n^u. The other conditions are clear from the definition.
(CG3) Again by (b$_n$), we have

$$\min\{e^{\varepsilon|k|}Q(\widehat{x}_{n+k}) : k \geq 0\} = e^{\pm\varepsilon/3}\min\{e^{\varepsilon|k|}Q(\widehat{f^{n+k}}(\widehat{x})) : k \geq 0\}$$

hence $\frac{p_n^s}{q^s(\widehat{f^n}(\widehat{x}))} = e^{\pm\varepsilon/3}$, and analogously $\frac{p_n^u}{q^u(\widehat{f^n}(\widehat{x}))} = e^{\pm\varepsilon/3}$. By Lemma 3.5(1), $p_n^s \wedge p_n^u = e^{\pm\varepsilon/3}q(\widehat{f^n}(\widehat{x})) \in [e^{-j_n-2}, e^{-j_n+2})$.

Claim 3: $\Psi_{\widehat{x}_n}^{p_n^s,p_n^u} \overset{\varepsilon}{\to} \Psi_{\widehat{x}_{n+1}}^{p_{n+1}^s,p_{n+1}^u}$ for all $n \in \mathbb{Z}$.

(GPO1) We have $\vartheta_1[\widehat{x}_n], \vartheta_0[\widehat{x}_{n+1}] \in \mathfrak{B}_{a_{n+1}}$, and by ($\mathrm{a}_n$) with $i = 1$ and (a_{n+1}) with $i = 0$, we have

$$\begin{aligned}
&d(\vartheta_1[\widehat{x}_n], \vartheta_0[\widehat{x}_{n+1}]) + \left\| \widetilde{C(\widehat{f}(\widehat{x}_n))} - \widetilde{C(\widehat{x}_{n+1})} \right\| \\
&\quad \le d(\vartheta_1[\widehat{f}^n(\widehat{x})], \vartheta_1[\widehat{x}_n]) + \left\| \widetilde{C(\widehat{f}^{n+1}(\widehat{x}))} - \widetilde{C(\widehat{f}(\widehat{x}_n))} \right\| \\
&\qquad + d(\vartheta_1[\widehat{f}^n(\widehat{x})], \vartheta_0[\widehat{x}_{n+1}]) + \left\| \widetilde{C(\widehat{f}^{n+1}(\widehat{x}))} - \widetilde{C(\widehat{x}_{n+1})} \right\| \\
&\quad < e^{-8(j_n+2)} + e^{-8(j_{n+1}+2)} \le e^{-8} \left(q(\widehat{f}^n(\widehat{x}))^8 + q(\widehat{f}^{n+1}(\widehat{x}))^8 \right) \\
&\quad \overset{!}{\le} e^{-8}(1 + e^{8\varepsilon}) q(\widehat{f}^{n+1}(\widehat{x}))^8 \\
&\quad \le e^{-8+8\varepsilon/3}(1 + e^{8\varepsilon})(p^s_{n+1} \wedge p^u_{n+1})^8 \overset{!!}{<} (p^s_{n+1} \wedge p^u_{n+1})^8,
\end{aligned}$$

where in $\overset{!}{\le}$ we used Lemma 3.4 and in $\overset{!!}{<}$ we used that $e^{-8+8\varepsilon/3}(1+e^{8\varepsilon}) < 1$ when $\varepsilon > 0$ is sufficiently small. This proves that $\Psi^{p^s_{n+1} \wedge p^u_{n+1}}_{\widehat{f}(\widehat{x}_n)} \overset{\varepsilon}{\approx} \Psi^{p^s_{n+1} \wedge p^u_{n+1}}_{\widehat{x}_{n+1}}$. Similarly, we prove that $\Psi^{p^s_n \wedge p^u_n}_{\widehat{f}^{-1}(\widehat{x}_{n+1})} \overset{\varepsilon}{\approx} \Psi^{p^s_n \wedge p^u_n}_{\widehat{x}_n}$.

(GPO2) The very definitions of p^s_n, p^u_n guarantee that $p^s_n = \min\{e^\varepsilon p^s_{n+1}, \delta_\varepsilon Q(\widehat{x}_n)\}$ and $p^u_{n+1} = \min\{e^\varepsilon p^u_n, \delta_\varepsilon Q(\widehat{x}_{n+1})\}$.

Claim 4: $\{\Psi^{p^s_n, p^u_n}_{\widehat{x}_n}\}_{n\in\mathbb{Z}}$ shadows $\widehat{x}$.

By (a_n) with $i = 0$, we have $\Psi^{p^s_n \wedge p^u_n}_{\widehat{f}^n(\widehat{x})} \overset{\varepsilon}{\approx} \Psi^{p^s_n \wedge p^u_n}_{\widehat{x}_n}$, hence by Proposition 4.1(4) we have $\vartheta_n[\widehat{x}] = \Psi_{\widehat{f}^n(\widehat{x})}(0) \in \Psi_{\widehat{x}_n}(B[p^s_n \wedge p^u_n])$, thus $\{\Psi^{p^s_n, p^u_n}_{\widehat{x}_n}\}_{n\in\mathbb{Z}}$ shadows $\widehat{x}$.

We thus proved that every $\widehat{x} \in \mathrm{NUH}^\#$ is shadowed by an ε-gpo $\{\Psi^{p^s_n, p^u_n}_{\widehat{x}_n}\}_{n\in\mathbb{Z}} \in \mathscr{A}^{\mathbb{Z}}$ s.t. $p^s_n \wedge p^u_n = e^{\pm\varepsilon/3} q(\widehat{f}^n(\widehat{x}))$. Since we have $\limsup_{n\to+\infty}(p^s_n \wedge p^u_n) > 0$ and $\limsup_{n\to-\infty}(p^s_n \wedge p^u_n) > 0$, the discreteness of $\mathscr{A}$ implies that $\{\Psi^{p^s_n, p^u_n}_{\widehat{x}_n}\}_{n\in\mathbb{Z}}$ is recurrent. This completes the proof of sufficiency.

Proof of relevance. The alphabet $\mathscr{A}$ might not a priori satisfy the relevance condition, but we can easily reduce it to a sub-alphabet $\mathscr{A}'$ satisfying (1)–(3). Call $v \in \mathscr{A}$ relevant if there is $\underline{v} \in \mathscr{A}^{\mathbb{Z}}$ with $v_0 = v$ s.t. $\underline{v}$ shadows a point in $\mathrm{NUH}^\#$. Since $\mathrm{NUH}^\#$ is $\widehat{f}$-invariant, every v_i is relevant. Then $\mathscr{A}' = \{v \in \mathscr{A} : v \text{ is relevant}\}$ is discrete because $\mathscr{A}' \subset \mathscr{A}$, it is sufficient and relevant by definition. □

Let Σ be the TMS associated to the graph with vertex set $\mathscr{A}$ given by Theorem 5.1 and edges $v \overset{\varepsilon}{\to} w$. An element of Σ is an ε-gpo, hence we define $\pi : \Sigma \to \widehat{M}$ by

$$\{\pi[\{v_n\}_{n\in\mathbb{Z}}]\} := \widehat{V}^s[\{v_n\}_{n\ge 0}] \cap \widehat{V}^u[\{v_n\}_{n\le 0}].$$

Recall the definition of $\Sigma^\#$. Here are the main properties of the triple (Σ, σ, π).

Proposition 5.2. *The following holds for all $\varepsilon > 0$ small enough.*

(1) *Each $v \in \mathscr{A}$ has finite ingoing and outgoing degree, hence Σ is locally compact.*
(2) *$\pi : \Sigma \to \widehat{M}$ is Hölder continuous.*

(3) $\pi \circ \sigma = \widehat{f} \circ \pi$.
(4) $\pi[\Sigma^\#] \supset \mathrm{NUH}^\#$.

PROOF. Part (1) follows from (GPO2), part (3) is obvious, and part (4) follows from Theorem 5.1(2). We prove part (2). Let $K > 1, \theta \in (0,1)$ satisfying Proposition 4.7(5). We will define constants $K_0 > 0$ and $\theta_0 \in (0,1)$ s.t. if $\underline{v} = \{v_n\}_{n\in\mathbb{Z}}, \underline{w} = \{w_n\}_{n\in\mathbb{Z}} \in \Sigma$ satisfy $v_n = w_n$ for $|n| \leq N$ then $d(\pi(\underline{v}), \pi(\underline{w})) \leq K_0\theta_0^N$. Recall that M has diameter smaller than one. Let $\theta_0 := \max\{\theta, \frac{1}{2}\} \in (0,1)$. For $\underline{v}, \underline{w}$ as above, write $\pi(\underline{v}) = \widehat{x} = \{x_n\}_{n\in\mathbb{Z}}$ and $\pi(\underline{w}) = \widehat{y} = \{y_n\}_{n\in\mathbb{Z}}$. By part (3), we have $\widehat{f}^n(\widehat{x}) = \pi[\sigma^n(\underline{v})]$ and $\widehat{f}^n(\widehat{y}) = \pi[\sigma^n(\underline{w})]$ and so $\{x_n\} = V^s[\sigma^n(\underline{v})] \cap V^u[\sigma^n(\underline{v})]$ and $\{y_n\} = V^s[\sigma^n(\underline{w})] \cap V^u[\sigma^n(\underline{w})]$. Fix $-N \leq n \leq 0$. Since $\sigma^n(\underline{v}), \sigma^n(\underline{w})$ coincide in all positions $-(N+n), \ldots, N+n$, Proposition 4.7(5) implies that $d_{C^1}(V^s[\sigma^n(\underline{v})], V^s[\sigma^n(\underline{w})]) \leq K\theta^{N+n}$ and $d_{C^1}(V^u[\sigma^n(\underline{v})], V^u[\sigma^n(\underline{w})]) \leq K\theta^{N+n}$. By Lemma 4.4, it follows that $d(x_n, y_n) \leq 6K\theta^{N+n}$ and so

$$2^n d(x_n, y_n) \leq 6K2^n\theta^{N+n} \leq 6K\theta_0^{-n}\theta_0^{N+n} = 6K\theta_0^N.$$

Noting that if $n < -N$ then $2^n d(x_n, y_n) < 2^n \leq \theta_0^{-n} < \theta_0^N$, we conclude that $d(\widehat{x}, \widehat{y}) \leq 6K\theta_0^N$. Letting $K_0 := 6K$, the proof is complete. □

It is important noting that (Σ, σ, π) does *not* satisfy Theorem 1.8, since π might be (and usually is) infinite-to-one. We use π to induce a locally finite cover of $\mathrm{NUH}^\#$, which will then be refined to a cover of $\mathrm{NUH}^\#$ by pairwise disjoint sets that will give a proof of Theorem 1.8.

CHAPTER 6

The inverse problem

Now we start investigating the non-injectivity of π. More specifically, if $\underline{v} = \{v_n\}_{n\in\mathbb{Z}}$ and $\underline{w} = \{w_n\}_{n\in\mathbb{Z}}$ satisfy $\pi(\underline{v}) = \pi(\underline{w})$, then we want to compare v_n and w_n and show that they are uniquely defined "up to bounded error". We do this under the additional assumption that $\underline{v}, \underline{w} \in \Sigma^\#$. Remind that $\Sigma^\#$ is the *recurrent set* of Σ:

$$\Sigma^\# := \left\{\underline{v} \in \Sigma : \exists v, w \in V \text{ s.t. } \begin{array}{l} v_n = v \text{ for infinitely many } n > 0 \\ v_n = w \text{ for infinitely many } n < 0 \end{array}\right\}.$$

The main result is the following.

THEOREM 6.1 (Inverse theorem). *The following holds for $\varepsilon > 0$ small enough. If $\{\Psi_{\widehat{x}_n}^{p_n^s,p_n^u}\}_{n\in\mathbb{Z}} \in \Sigma^\#$ satisfies $\pi[\{\Psi_{\widehat{x}_n}^{p_n^s,p_n^u}\}_{n\in\mathbb{Z}}] = \widehat{x}$ then $\widehat{x} \in \mathrm{NUH}^*$, and for all $n \in \mathbb{Z}$:*

(1) $d(\vartheta[\widehat{x}_n], \vartheta[\widehat{f}^n(\widehat{x})]) < 25^{-1}(p_n^s \wedge p_n^u)$ *and* $\frac{\rho(\widehat{x}_n)}{\rho(\widehat{f}^n(\widehat{x}))} = e^{\pm\varepsilon}$.

(2) $\frac{\|C(\widehat{x}_n)^{-1}\|}{\|C(\widehat{f}^n(\widehat{x}))^{-1}\|} = e^{\pm 2\sqrt{\varepsilon}}$.

(3) $\frac{Q(\widehat{x}_n)}{Q(\widehat{f}^n(\widehat{x}))} = e^{\pm\sqrt[3]{\varepsilon}}$.

(4) $\frac{p_n^s}{q^s(\widehat{f}^n(\widehat{x}))} = e^{\pm\sqrt[3]{\varepsilon}}$ *and* $\frac{p_n^u}{q^u(\widehat{f}^n(\widehat{x}))} = e^{\pm\sqrt[3]{\varepsilon}}$.

(5) $(\Psi_{\widehat{f}^n(\widehat{x})}^{-1} \circ \Psi_{\widehat{x}_n})(v) = \delta_n + O_n v + \Delta_n(v)$ *for $v \in B[10Q(\widehat{x}_n)]$, where $\delta_n \in \mathbb{R}^m$ satisfies $\|\delta_n\| < 25^{-1}(p_n^s \wedge p_n^u)$, O_n is an orthogonal linear map preserving the splitting $\mathbb{R}^d \times \mathbb{R}^{m-d}$, and $\Delta_n : B[10Q(\widehat{x}_n)] \to \mathbb{R}^m$ satisfies $\Delta_n(0) = 0$ and $\|d\Delta_n\|_{C^0} < 5\sqrt{\varepsilon}$ on $B[10Q(\widehat{x}_n)]$.*

In particular, $\widehat{x} \in \mathrm{NUH}^\#$.

The above theorem compares the parameters of an ε-gpo with the parameters of its shadowed point, and it allows to identify the coded set in Theorem 1.8. Below, we state a corollary that is useful for comparing the parameters of two ε-gpo's that shadow a same point.

COROLLARY 6.2. *The following holds for $\varepsilon > 0$ small enough. If $\underline{v} = \{\Psi_{\widehat{x}_n}^{p_n^s,p_n^u}\}_{n\in\mathbb{Z}} \in \Sigma^\#$ and $\underline{w} = \{\Psi_{\widehat{y}_n}^{q_n^s,q_n^u}\}_{n\in\mathbb{Z}} \in \Sigma^\#$ satisfy $\pi(\underline{v}) = \pi(\underline{w})$, then for all $n \in \mathbb{Z}$:*

(1) $d(\vartheta[\widehat{x}_n], \vartheta[\widehat{y}_n]) < 10^{-1}\max\{p_n^s \wedge p_n^u, q_n^s \wedge q_n^u\}$.

(2) $\frac{p_n^s}{q_n^s} = e^{\pm 2\sqrt[3]{\varepsilon}}$ *and* $\frac{p_n^u}{q_n^u} = e^{\pm 2\sqrt[3]{\varepsilon}}$.

(3) $(\Psi_{\widehat{y}_n}^{-1} \circ \Psi_{\widehat{x}_n})(v) = \delta_n + O_n v + \Delta_n(v)$ *for $v \in B[10Q(\widehat{x}_n)]$, where $\delta_n \in \mathbb{R}^m$ satisfies $\|\delta_n\| < 10^{-1}(q_n^s \wedge q_n^u)$, O_n is an orthogonal linear map preserving the splitting $\mathbb{R}^d \times \mathbb{R}^{m-d}$, and $\Delta_n : B[10Q(\widehat{x}_n)] \to \mathbb{R}^m$ satisfies $\Delta_n(0) = 0$ and $\|d\Delta_n\|_{C^0} < 9\sqrt{\varepsilon}$ on $B[10Q(\widehat{x}_n)]$.*

In the following section, we provide some preliminary results that will be used to prove Theorem 6.1. After this preparation is done, we will proceed to prove each part of 6.1. Clearly, we only need to prove it for $n = 0$.

6.1. Identification of invariant subspaces

In order to analyse the inverse problem, we have to compare the parameters of charts. One of the steps is to compare the hyperbolicity parameters, which are related to the Lyapunov inner product, see Section 3.1. In principle, for $\widehat{x}, \widehat{y} \in \mathrm{NUH}$ with $d(\vartheta[\widehat{x}], \vartheta[\widehat{y}]) \ll 1$, it is not clear how to compare the norm in the Lyapunov inner product between vectors in $E^{s/u}_{\widehat{x}}$ and in $E^{s/u}_{\widehat{y}}$. When $\widehat{x}$ defines an ε-double chart $\Psi^{p^s,p^u}_{\widehat{x}}$ and $\widehat{y}$ belongs to a stable/unstable set at $\Psi^{p^s,p^u}_{\widehat{x}}$, Ben Ovadia proposed a canonical way of making this comparison, using the representing function in the chart [**BO18**]. The idea is to make an identification of subspaces in the chart and then push it to the manifold. Here, we follow the same approach, adapted to our context. Since the unstable subspace depends on inverse branches, the definition is made inside $\widehat{TM}$. In particular, if $\widehat{y}$ is the intersection of a stable and an unstable set, both at an ε-double chart $\Psi^{p^s,p^u}_{\widehat{x}}$, then we can compare the tangent spaces $\widehat{TM}_{\widehat{x}}$ and $\widehat{TM}_{\widehat{y}}$.

Let $\Psi^{p^s,p^u}_{\widehat{x}}$ be an ε-double chart, let $V = V^s$ be an s-admissible manifold at $\Psi^{p^s,p^u}_{\widehat{x}}$ and $\widehat{V}$ be the respective stable set. In the sequel, we use again the notation $T\widehat{V} = \widehat{TV}$ to represent the tangent bundle of V as a subset of $\widehat{TM}$. Writing $d = d_s(\widehat{x})$, recall that

$$V = \Psi_{\widehat{x}}\{(v_1, G(v_1)) : v_1 \in B^d[p^s]\}$$

where $G : B^d[p^s] \to \mathbb{R}^{m-d}$ is a $C^{1+\beta/3}$ function satisfying (AM1)–(AM3). Fix $\widehat{y} \in \widehat{V}$, and write $\vartheta[\widehat{y}] = y = \Psi_{\widehat{x}}(z)$ where $z = (v_1, G(v_1))$. Denote the tangent space to the graph of G at z by $T_z \operatorname{Graph}(G)$. We have

$$T_z \operatorname{Graph}(G) = \left\{ \begin{bmatrix} w \\ (dG)_{v_1} w \end{bmatrix} : w \in \mathbb{R}^d \right\},$$

which is canonically isomorphic to $\mathbb{R}^d \times \{0\}$ via the map

$$\begin{aligned} \iota_s \ : \ \mathbb{R}^d \times \{0\} &\longrightarrow T_z \operatorname{Graph}(G) \\ \begin{bmatrix} w \\ 0 \end{bmatrix} &\longmapsto \begin{bmatrix} w \\ (dG)_{v_1} w \end{bmatrix}. \end{aligned}$$

Recalling that $E^s_{\widehat{x}} = (d\Psi_{\widehat{x}})_0[\mathbb{R}^d \times \{0\}]$ and $T_{\widehat{y}}\widehat{V} = (d\Psi_{\widehat{x}})_z[T_z \operatorname{Graph}(G)]$, we have the following definition.

THE MAP $\Theta^s_{\widehat{x},\widehat{y}}$: We define $\Theta^s_{\widehat{x},\widehat{y}} : E^s_{\widehat{x}} \to T_{\widehat{y}}\widehat{V}$ as the composition of the linear maps

$$\Theta^s_{\widehat{x},\widehat{y}} := (d\Psi_{\widehat{x}})_z \circ \iota_s \circ [(d\Psi_{\widehat{x}})_0]^{-1}.$$

In other words, $\Theta^s_{\widehat{x},\widehat{y}}$ is defined to make the diagram below commute

$$\begin{array}{ccc} \mathbb{R}^d \times \{0\} & \xrightarrow{\iota_s} & T_z\,\mathrm{Graph}(G) \\ \downarrow{\scriptstyle (d\Psi_{\widehat{x}})_0} & & \downarrow{\scriptstyle (d\Psi_{\widehat{x}})_z} \\ E^s_{\widehat{x}} & \xrightarrow[\Theta^s_{\widehat{x},\widehat{y}}]{} & T_{\widehat{y}}\widehat{V} \end{array}$$

so that

$$\Theta^s_{\widehat{x},\widehat{y}}\left((d\Psi_{\widehat{x}})_0\begin{bmatrix} w \\ 0 \end{bmatrix}\right) = (d\Psi_{\widehat{x}})_z\begin{bmatrix} w \\ (dG)_{v_1}w \end{bmatrix}.$$

A similar definition holds when $\widehat{y}$ belongs to an unstable set at $\Psi^{p^s,p^u}_{\widehat{x}}$. Let

$$V = \Psi_{\widehat{x}}\{(G(v_2), v_2) : v_2 \in B^{m-d}[p^u]\}$$

where $G : B^{m-d}[p^u] \to \mathbb{R}^d$ is a $C^{1+\beta/3}$ function satisfying (AM1)–(AM3), and let $\widehat{V}$ be the respective unstable set. For $\widehat{y} \in \widehat{V}$, write $\vartheta[\widehat{y}] = y = \Psi_{\widehat{x}}(z)$ where $z = (G(v_2), v_2)$. Now

$$T_z\,\mathrm{Graph}(G) = \left\{\begin{bmatrix} (dG)_{v_2}w \\ w \end{bmatrix} : w \in \mathbb{R}^{m-d}\right\},$$

which is canonically isomorphic to $\{0\} \times \mathbb{R}^{m-d}$ via the map

$$\begin{array}{rcccl} \iota_u & : & \{0\} \times \mathbb{R}^{m-d} & \longrightarrow & T_z\,\mathrm{Graph}(G) \\ & & \begin{bmatrix} 0 \\ w \end{bmatrix} & \longmapsto & \begin{bmatrix} (dG)_{v_2}w \\ w \end{bmatrix}. \end{array}$$

THE MAP $\Theta^u_{\widehat{x},\widehat{y}}$: We define $\Theta^u_{\widehat{x},\widehat{y}} : E^u_{\widehat{x}} \to T_{\widehat{y}}\widehat{V}$ as the composition of the linear maps

$$\Theta^u_{\widehat{x},\widehat{y}} := (d\Psi_{\widehat{x}})_z \circ \iota_u \circ [(d\Psi_{\widehat{x}})_0]^{-1}.$$

Similarly, $\Theta^u_{\widehat{x},\widehat{y}}$ is defined to make the following diagram to commute

$$\begin{array}{ccc} \{0\} \times \mathbb{R}^{m-d} & \xrightarrow{\iota_u} & T_z\,\mathrm{Graph}(G) \\ \downarrow{\scriptstyle (d\Psi_{\widehat{x}})_0} & & \downarrow{\scriptstyle (d\Psi_{\widehat{x}})_z} \\ E^u_{\widehat{x}} & \xrightarrow[\Theta^u_{\widehat{x},\widehat{y}}]{} & T_{\widehat{y}}\widehat{V} \end{array}$$

and so

$$\Theta^u_{\widehat{x},\widehat{y}}\left((d\Psi_{\widehat{x}})_0\begin{bmatrix} 0 \\ w \end{bmatrix}\right) = (d\Psi_{\widehat{x}})_z\begin{bmatrix} (dG)_{v_2}w \\ w \end{bmatrix}.$$

Now assume that $\{\widehat{y}\} = \widehat{V}^s \cap \widehat{V}^u$, where $\widehat{V}^{s/u}$ is a stable/unstable set at $\Psi^{p^s,p^u}_{\widehat{x}}$. By the above discussion, we have two linear maps $\Theta^{s/u}_{\widehat{x},\widehat{y}} : E^{s/u}_{\widehat{x}} \to T_{\widehat{y}}\widehat{V}^{s/u}$. Since $E^s_{\widehat{x}} \oplus E^u_{\widehat{x}} = \widehat{TM}_{\widehat{x}}$ and $T_{\widehat{y}}\widehat{V}^s \oplus T_{\widehat{y}}\widehat{V}^u = \widehat{TM}_{\widehat{y}}$, the following definition makes sense.

THE MAP $\Theta_{\widehat{x},\widehat{y}}$: We define $\Theta_{\widehat{x},\widehat{y}} : \widehat{TM}_{\widehat{x}} \to \widehat{TM}_{\widehat{y}}$ as the unique linear map s.t. $\Theta_{\widehat{x},\widehat{y}} \restriction_{E^s_{\widehat{x}}} = \Theta^s_{\widehat{x},\widehat{y}}$ and $\Theta_{\widehat{x},\widehat{y}} \restriction_{E^u_{\widehat{x}}} = \Theta^u_{\widehat{x},\widehat{y}}$. More specifically, if $v = v^s + v^u$ with $v^{s/u} \in E^{s/u}_{\widehat{x}}$ then

$$\Theta_{\widehat{x},\widehat{y}}(v) := \Theta^s_{\widehat{x},\widehat{y}}(v^s) + \Theta^u_{\widehat{x},\widehat{y}}(v^u).$$

We can similarly see $\Theta_{\widehat{x},\widehat{y}}$ as a map defined in terms of a commuting diagram. Let G, H be the representing functions of $\widehat{V}^s, \widehat{V}^u$, let $\vartheta[\widehat{y}] = y = \Psi_{\widehat{x}}(z)$ with $z = (v_1, v_2)$, and let

$$\begin{aligned}\iota : \mathbb{R}^m &\longrightarrow \mathbb{R}^m \\ \begin{bmatrix} w_1 \\ w_2 \end{bmatrix} &\longmapsto \begin{bmatrix} w_1 + (dH)_{v_2} w_2 \\ w_2 + (dG)_{v_1} w_1 \end{bmatrix}.\end{aligned}$$

Then we obtain a commuting diagram

$$\begin{array}{ccc} \mathbb{R}^m & \xrightarrow{\quad \iota \quad} & \mathbb{R}^m \\ {\scriptstyle (d\Psi_{\widehat{x}})_0} \big\downarrow & & \big\downarrow {\scriptstyle (d\Psi_{\widehat{x}})_z} \\ \widehat{TM}_{\widehat{x}} & \xrightarrow[\Theta_{\widehat{x},\widehat{y}}]{} & \widehat{TM}_{\widehat{y}} \end{array}$$

and so

$$\Theta_{\widehat{x},\widehat{y}}\left((d\Psi_{\widehat{x}})_0 \begin{bmatrix} w_1 \\ w_2 \end{bmatrix} \right) = (d\Psi_{\widehat{x}})_z \begin{bmatrix} w_1 + (dH)_{v_2} w_2 \\ w_2 + (dG)_{v_1} w_1 \end{bmatrix}.$$

The next lemma proves that the maps just defined are close to parallel transports.

LEMMA 6.3. *Let $\Psi_{\widehat{x}}^{p^s,p^u}$ be an ε-double chart with $\vartheta[\widehat{x}] = x$ and $\eta = p^s \wedge p^u$.*

(1) *Let $\widehat{V}$ be a stable set at $\Psi_{\widehat{x}}^{p^s,p^u}$, and let $\widehat{y} \in \widehat{V}$. If $\vartheta[\widehat{y}] = y \in \Psi_{\widehat{x}}(B^d[\eta] \times B^{m-d}[\eta])$ then*

$$\left\| \Theta^s_{\widehat{x},\widehat{y}} - P_{x,y} \right\| \le \tfrac{1}{2}\eta^{15\beta/48},$$

where $P_{x,y}$ is the restriction of the parallel transport from x to y to the subspace $E^s_{\widehat{x}}$. In particular, $\frac{\|\Theta^s_{\widehat{x},\widehat{y}}(v)\|}{\|v\|} = \exp\left[\pm\eta^{15\beta/48}\right]$ for all $v \in E^s_{\widehat{x}}$. An analogous statement holds for unstable sets.

(2) *If $\widehat{y} \in \widehat{V}^s \cap \widehat{V}^u$ where $\widehat{V}^{s/u}$ is a stable/unstable set at $\Psi_{\widehat{x}}^{p^s,p^u}$, then*

$$\left\| \Theta_{\widehat{x},\widehat{y}} - P_{x,y} \right\| \le \tfrac{1}{2}\eta^{15\beta/48}.$$

In particular, $\frac{\|\Theta_{\widehat{x},\widehat{y}}(v)\|}{\|v\|} = \exp\left[\pm\eta^{15\beta/48}\right]$ for all $v \in \widehat{TM}_{\widehat{x}}$.

PROOF. We start proving part (2). For simplicity, write $\Theta = \Theta_{\widehat{x},\widehat{y}}$ and $P = P_{x,y}$. By Lemma 4.4, $\|v_1\| \le \|(v_1, v_2)\| < \eta$. By (AM2)–(AM3),

$$\|(dG)_{v_1}\| \le \|(dG)_0\| + \tfrac{1}{2}\|v_1\|^{\beta/3} \le \tfrac{1}{2}\eta^{\beta/3} + \tfrac{1}{2}\eta^{\beta/3} = \eta^{\beta/3}.$$

Similarly, $\|(dH)_{v_2}\| \le \eta^{\beta/3}$. Hence $\iota = \mathrm{Id} + L$, where $\|L\| \le \eta^{\beta/3}$. Writing $y = \Psi_{\widehat{x}}(z)$ for $z \in B^d[\eta] \times B^{m-d}[\eta]$, we have $\Theta = (d\Psi_{\widehat{x}})_z \circ \iota \circ [(d\Psi_{\widehat{x}})_0]^{-1}$ and so

$$\Theta = (d\Psi_{\widehat{x}})_z \circ [(d\Psi_{\widehat{x}})_0]^{-1} + E = (d\exp_x)_{C(\widehat{x})z} + E$$

where $E = (d\exp_x)_{C(\widehat{x})z} \circ C(\widehat{x}) \circ L \circ C(\widehat{x})^{-1}$. Therefore $\Theta - P = (d\exp_x)_{C(\widehat{x})z} - P + E$. Observe that:

- $(d\exp_x)_{C(\widehat{x})z} - P = \widetilde{(d\exp_x)_{C(\widehat{x})z}} - \widetilde{(d\exp_x)_0}$ and so by (A3) we have

$$\left\|(d\exp_x)_{C(\widehat{x})z} - P\right\| \le d(x, \mathscr{S})^{-a}\|z\| \le 2\rho(\widehat{x})^{-a}\eta \le 2\varepsilon^{1/16}\eta^{95\beta/96},$$

where in the last passage we used that $\rho(\widehat{x})^{-a}\eta^{\beta/96} \le \rho(\widehat{x})^{-a}Q(\widehat{x})^{\beta/96} \le \varepsilon^{1/16}$.

- $\|E\| \le 2\|C(\widehat{x})^{-1}\|\|L\| \le 2\|C(\widehat{x})^{-1}\|\eta^{\beta/3} \le 2\varepsilon^{1/8}\eta^{15\beta/48}$, where in the last passage we used that $\|C(\widehat{x})^{-1}\|\eta^{\beta/48} \le \|C(\widehat{x})^{-1}\|Q(\widehat{x})^{\beta/48} \le \varepsilon^{1/8}$.

Therefore, if $\varepsilon > 0$ is small enough then

$$\|\Theta - P\| \le 2\varepsilon^{1/16}\eta^{95\beta/96} + 2\varepsilon^{1/8}\eta^{15\beta/48} < \tfrac{1}{2}\eta^{15\beta/48}.$$

This proves part (2). Part (1) is proved exactly in the same manner, using that $\iota_s = \mathrm{Id}_{\mathbb{R}^d \times \{0\}} + L$ and that the same estimates above hold for L. □

Compare Lemma 6.3 with the estimates in the proof of [**BO18**, Lemma 4.6]. Our assumption is stronger because we require y to be very close to the center of the chart $\Psi_{\widehat{x}}^{p^s,p^u}$, but we obtain a much better estimate. This sharpened estimate will be used in the next section, for the following reason. Given an edge $\Psi_{\widehat{x}}^{p^s,p^u} \overset{\varepsilon}{\to} \Psi_{\widehat{y}}^{q^s,q^u}$, we will need to compare the parameters of the two charts. The usual implementation, under the assumption that f is a diffeomorphism, makes this comparison using that the ratio $\frac{Q(\widehat{f}(\widehat{x}))}{Q(\widehat{x})}$ is bounded, since the derivative is uniformly bounded. This is not the case in our context, and there is no a priori bound on the ratio $\frac{Q(\widehat{f}(\widehat{x}))}{Q(\widehat{x})}$. On the other hand, we know that $\frac{p^s \wedge p^u}{q^s \wedge q^u} = e^{\pm\varepsilon}$. Hence, provided our estimates depend on $p^s \wedge p^u$ and $q^s \wedge q^u$, we will be able to compare the parameters of $\Psi_{\widehat{x}}^{p^s,p^u}$ with those of $\Psi_{\widehat{y}}^{q^s,q^u}$.

6.2. Improvement lemma

As we have seen in Lemma 3.1, the Lyapunov inner product allows to "transform" a nonuniformly hyperbolic system into a uniformly hyperbolic one and then use the classical tools of uniform hyperbolicity, such as graph transforms. One essential feature of uniform hyperbolicity is the following: forward iterations of the system improve the behavior along unstable directions (the forward expansion uniformizes the behavior), and backward iterations improve the behavior along stable directions. This philosophy is behind the recent work on anisotropic spaces, see e.g. [**BKL02**]. The main result of this section is another manifestation of this behavior. We call it an *improvement lemma*, as in [**LM18**]. Recall the functions S, U introduced in (NUH3) and the graph transforms $\widehat{\mathscr{F}}^{s/u}$ introduced in Section 4.6. Throughout this section, d is the dimension of the stable subspace.

LEMMA 6.4 (Improvement lemma). *The following holds for all $\varepsilon > 0$ small enough. Let $v \overset{\varepsilon}{\to} w$ with $v = \Psi_{\widehat{x}_0}^{p_0^s,p_0^u}$, $w = \Psi_{\widehat{x}_1}^{p_1^s,p_1^u}$, and write $\eta_0 = p_0^s \wedge p_0^u$, $\eta_1 = p_1^s \wedge p_1^u$. Fix $\widehat{V}_1 \in \widehat{\mathscr{M}}^s(w)$ and let $\widehat{V}_0 = \widehat{\mathscr{F}}^s_{v,w}[\widehat{V}_1]$. Let $\widehat{y}_0 \in \widehat{V}_0$, $\widehat{y}_1 = \widehat{f}(\widehat{y}_0) \in \widehat{V}_1$, and assume that*

$$\vartheta[\widehat{y}_0] \in \Psi_{\widehat{x}_0}(B^d[\eta_0] \times B^{m-d}[\eta_0]) \quad \textit{and} \quad \vartheta[\hat{y}_1] \in \Psi_{\widehat{x}_1}(B^d[\eta_1] \times B^{m-d}[\eta_1]).$$

Let $v_1 \in T_{\widehat{y}_1}\widehat{V}_1$ and $v_0 = \widehat{df}^{(-1)} v_1 \in T_{\widehat{y}_0}\widehat{V}_0$. Finally, let $w_0 \in \widehat{TM}_{\widehat{x}_0}$ s.t. $v_0 = \Theta^s_{\widehat{x}_0,\widehat{y}_0}(w_0)$ and $w_1 \in \widehat{TM}_{\widehat{x}_1}$ s.t. $v_1 = \Theta^s_{\widehat{x}_1,\widehat{y}_1}(w_1)$. If $\frac{S(\widehat{y}_1,v_1)}{S(\widehat{x}_1,w_1)} = \exp[\pm\xi]$ for $\xi \ge \sqrt{\varepsilon}$, then

$$\frac{S(\widehat{y}_0, v_0)}{S(\widehat{x}_0, w_0)} = \exp\left[\pm(\xi - Q(\widehat{x}_0)^{\beta/4})\right].$$

An analogous statement holds for unstable sets.

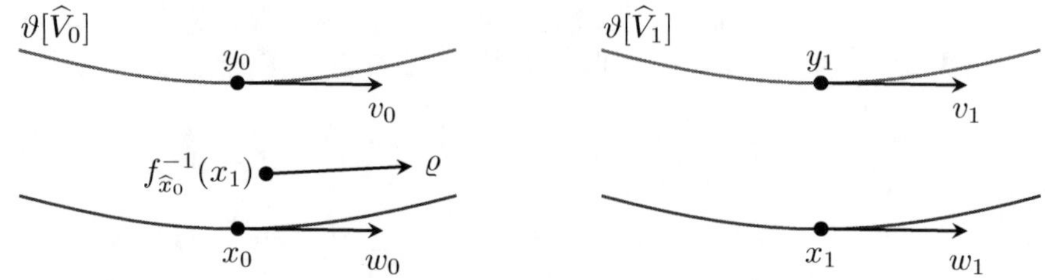

FIGURE 6.1. Points and vectors in the improvement lemma.

PROOF. Let $x_0 = \vartheta[\widehat{x}_0]$, $y_0 = \vartheta[\widehat{y}_0]$, $x_1 = \vartheta[\widehat{x}_1]$, $y_1 = \vartheta[\widehat{y}_1]$. Also let $\varrho = \widehat{df}^{(-1)} w_1 = (df_{\widehat{x}_0}^{-1})_{x_1} w_1 \in \widehat{TM}_{\widehat{f}^{-1}(\widehat{x}_1)}$, see Figure 6.1. In the following, the base-points in $\widehat{TM}$ of all vectors involved in the calculations are fixed, hence we write $S(\cdot, v)$ simply by $|||v|||$. We focus on one side of the estimate: under the assumption that $\frac{|||v_1|||}{|||w_1|||} \leq \exp[\xi]$ for $\xi \geq \sqrt{\varepsilon}$ we will prove that $\frac{|||v_0|||}{|||w_0|||} \leq \exp[\xi - Q(\widehat{x}_0)^{\beta/4}]$. The inequality in the other direction will follow in the same way. We are allowed to make one normalization, so we take $\|v_0\| = 1$. We write $\frac{|||v_0|||}{|||w_0|||} = \frac{|||v_0|||}{|||\varrho|||} \cdot \frac{|||\varrho|||}{|||w_0|||}$ and estimate each of the fractions separately. Before, we compare the norms of the fiber components of the vectors v_0, w_0, ϱ. The calculations are simple, once we pay attention to which parallel transports we apply.

- Comparison of v_0, w_0: by Lemma 6.3(1),

$$\|P_{y_0,x_0} v_0 - w_0\| \leq \tfrac{1}{2}\eta_0^{15\beta/48} < \tfrac{1}{2}\eta_0^{14\beta/48}. \tag{6.1}$$

- Comparison of v_0, ϱ: recalling the definitions of ϱ and v_0, we have

$$\begin{aligned}
\left\|\varrho - P_{y_0, f_{\widehat{x}_0}^{-1}(x_1)} v_0\right\| &= \left\|(df_{\widehat{x}_0}^{-1})_{x_1} w_1 - P_{y_0, f_{\widehat{x}_0}^{-1}(x_1)} \circ (df_{\widehat{x}_0}^{-1})_{y_1} v_1\right\| \\
&= \left\|(df_{\widehat{x}_0}^{-1})_{x_1} w_1 - \widetilde{(df_{\widehat{x}_0}^{-1})_{y_1}}[P_{y_1,x_1} v_1]\right\| \\
&\leq \left\|(df_{\widehat{x}_0}^{-1})_{x_1}\right\| \|w_1 - P_{y_1,x_1} v_1\| + \left\|\widetilde{(df_{\widehat{x}_0}^{-1})_{x_1}} - \widetilde{(df_{\widehat{x}_0}^{-1})_{y_1}}\right\| \|v_1\|.
\end{aligned}$$

By Lemma 6.3(1), $\|w_1 - P_{y_1,x_1} v_1\| \leq \frac{1}{2}\|v_1\| \eta_1^{15\beta/48} \leq d(x_0, \mathscr{S})^{-a} \eta_0^{15\beta/48}$ for $\varepsilon > 0$ small enough, since $\eta_1 \leq e^{\varepsilon}\eta_0$. Using assumptions (A6)–(A7) and that $d(x_1, y_1) \leq 2\eta_1 \leq 4\eta_0$, we conclude that

$$\begin{aligned}
\left\|\varrho - P_{y_0, f_{\widehat{x}_0}^{-1}(x_1)} v_0\right\| &\leq d(x_0, \mathscr{S})^{-2a} \eta_0^{15\beta/48} + \mathfrak{K}(4\eta_0)^{\beta} d(x_0, \mathscr{S})^{-a} \\
&\leq d(x_0, \mathscr{S})^{-2a} \eta_0^{15\beta/48} + 4\mathfrak{K} d(x_0, \mathscr{S})^{-a} \eta_0^{\beta} \\
&\leq \varepsilon^{1/8} \eta_0^{14\beta/48} + 4\varepsilon^{1/16} \mathfrak{K} \eta_0^{95\beta/96} \\
&< \tfrac{1}{2}\eta_0^{14\beta/48}.
\end{aligned}$$

- Comparison of w_0, ϱ: the two estimates above imply that

$$\begin{aligned}
\left\|\varrho - P_{x_0, f_{\widehat{x}_0}^{-1}(x_1)} w_0\right\| &\leq \left\|\varrho - P_{y_0, f_{\widehat{x}_0}^{-1}(x_1)} v_0\right\| + \left\|P_{y_0, f_{\widehat{x}_0}^{-1}(x_1)} v_0 - P_{x_0, f_{\widehat{x}_0}^{-1}(x_1)} w_0\right\| \\
&= \left\|\varrho - P_{y_0, f_{\widehat{x}_0}^{-1}(x_1)} v_0\right\| + \|P_{y_0,x_0} v_0 - w_0\| < \eta_0^{14\beta/48}.
\end{aligned}$$

Now we estimate the fractions $\frac{|||v_0|||}{|||\varrho|||}$ and $\frac{|||\varrho|||}{|||w_0|||}$.

ESTIMATE OF $\frac{|||\varrho|||}{|||w_0|||}$: by the definition of S and Lemma 6.3(1),

$$|||w_0||| = \left\|C(\widehat{x}_0)^{-1}w_0\right\| \geq \|w_0\| \geq \exp\left[-\eta_0^{15\beta/48}\right] \geq \tfrac{1}{2}.$$

Therefore, letting $A = C(\widehat{x}_0)^{-1}$ and $B = C(\widehat{f}^{-1}(\widehat{x}_1))^{-1}$, we have

$$\left|\frac{|||\varrho|||}{|||w_0|||} - 1\right| \leq 2\left||||\varrho||| - |||w_0|||\right| = 2\left|\|B\varrho\| - \|Aw_0\|\right|.$$

Let $\widetilde{B} = B \circ P_{x_0, f_{\widehat{x}_0}^{-1}(x_1)}$, so that $A, \widetilde{B}$ have the same domain (and codomain). Then

$$\begin{aligned}\left|\|B\varrho\| - \|Aw_0\|\right| &\leq \left|\|B\varrho\| - \|\widetilde{B}w_0\|\right| + \left\|\widetilde{B}w_0 - Aw_0\right\| \\ &\leq \left\|B\varrho - B \circ P_{x_0, f_{\widehat{x}_0}^{-1}(x_1)}w_0\right\| + \left\|\widetilde{B}w_0 - Aw_0\right\| \\ &\leq \|B\|\left\|\varrho - P_{x_0, f_{\widehat{x}_0}^{-1}(x_1)}w_0\right\| + \left\|\widetilde{B} - A\right\|\|w_0\|\end{aligned}$$

By the overlap condition, Proposition 4.1(2) and the comparison of w_0, ϱ, the latter expression above is at most $2\|A\|\eta_0^{14\beta/48} + 2\eta_0 \leq 2\varepsilon^{1/8}\eta_0^{13\beta/48} + 2\eta_0 < \frac{1}{4}\eta_0^{\beta/4}$. Hence $\left|\frac{|||\varrho|||}{|||w_0|||} - 1\right| < \frac{1}{2}\eta_0^{\beta/4}$ and so

$$\frac{|||\varrho|||}{|||w_0|||} = \exp\left[\pm\eta_0^{\beta/4}\right]. \tag{6.2}$$

ESTIMATE OF $\frac{|||v_0|||}{|||\varrho|||}$: by the latter estimate, we need to prove that

$$\frac{|||v_0|||^2}{|||\varrho|||^2} \leq \exp\left[2\xi - 4Q(\widehat{x}_0)^{\beta/4}\right]. \tag{6.3}$$

Recall our assumption that $\frac{|||v_1|||}{|||w_1|||} \leq e^{\xi}$ for $\xi \geq \sqrt{\varepsilon}$. By relation (3.1),

$$\begin{aligned}\frac{|||v_0|||^2}{|||\varrho|||^2} &= \frac{2\|v_0\|^2 + e^{2\chi}|||v_1|||^2}{2\|\varrho\|^2 + e^{2\chi}|||w_1|||^2} \\ &\leq \frac{2\|v_0\|^2 + e^{2\chi}e^{2\xi}|||w_1|||^2}{2\|\varrho\|^2 + e^{2\chi}|||w_1|||^2} \\ &= e^{2\xi}\left(\frac{\frac{2\|v_0\|^2}{e^{2\xi}} + e^{2\chi}|||w_1|||^2}{2\|\varrho\|^2 + e^{2\chi}|||w_1|||^2}\right) \\ &= e^{2\xi}\left(1 - \frac{2\|\varrho\|^2 - 2\frac{\|v_0\|^2}{e^{2\xi}}}{2\|\varrho\|^2 + e^{2\chi}|||w_1|||^2}\right) \\ &= e^{2\xi}\left(1 - 2\frac{\|\varrho\|^2 - \frac{\|v_0\|^2}{e^{2\xi}}}{|||\varrho|||^2}\right)\end{aligned}$$

hence (6.3) will follow once we prove that (remind that $\|v_0\| = 1$)

$$\frac{\|\varrho\|^2 - e^{-2\xi}}{|||\varrho|||^2} \geq 2Q(\widehat{x}_0)^{\beta/4} \iff \|\varrho\|^2 - e^{-2\xi} \geq 2|||\varrho|||^2 Q(\widehat{x}_0)^{\beta/4}.$$

We estimate the order of ε in both sides of this last inequality:

- By the comparison of v_0, ϱ, we have $|\|\varrho\| - 1| < \frac{1}{2}\eta_0^{14\beta/48} < Q(\widehat{x}_0)^{\beta/4} < \varepsilon^{3/2}$. Since $\xi \geq \sqrt{\varepsilon}$, we have $1 - e^{-2\xi} \geq 1 - e^{-2\sqrt{\varepsilon}} > \varepsilon^{1/2}$ and so

$$\|\varrho\|^2 - e^{-2\xi} = (1 - e^{-2\xi}) + \left(\|\varrho\|^2 - 1\right) \geq (1 - e^{-2\xi}) - \left|\|\varrho\|^2 - 1\right| > \varepsilon^{1/2} - 3\varepsilon^{3/2}.$$

- Since $\|\|\varrho\|\| = \|B\varrho\| \leq 2\|B\| \leq 4\|A\|$, we have

$$2\|\|\varrho\|\|^2 Q(\widehat{x}_0)^{\beta/4} \leq 32\|A\|^2 Q(\widehat{x}_0)^{\beta/4} \leq 32\varepsilon^{1/4} Q(\widehat{x}_0)^{5\beta/24} \leq 32\varepsilon^{1/4}\varepsilon^{5/4} = 32\varepsilon^{3/2}.$$

If $\varepsilon > 0$ is small enough then $\varepsilon^{1/2} - 3\varepsilon^{3/2} > 32\varepsilon^{3/2}$. This proves (6.3) which, together with (6.2), gives that $\frac{\|\|v_0\|\|}{\|\|w_0\|\|} \leq \exp[\xi - Q(\widehat{x})^{\beta/4}]$. The proof for stable sets is now complete.

The proof for unstable sets is the same, as we now explain. When dealing with the unstable direction, we usually need to identify which inverse branches are being used. So we would expect to require that, for the proof of the improvement lemma, the inverse branches associated with $\widehat{f}(\widehat{x}_0)$ and $\widehat{x}_1$ are the same, so that we can compare $U(\widehat{f}(\widehat{x}_0), \widehat{df}w_0)$ and $U(\widehat{x}_1, w_1)$. This is indeed *not* necessary, since this comparison comes directly from the overlap condition, which guarantees that these two values are almost the same! Therefore, the same proof applied for stable sets also works for unstable sets. □

As we just saw, the proof of the improvement lemma relies on the estimate of the fractions $\frac{\|\|\varrho\|\|}{\|\|w_0\|\|}$ and $\frac{\|\|v_0\|\|}{\|\|\varrho\|\|}$. The estimate of the first fraction is consequence of the overlap condition, which allows to compare the matrices A and B. The overlap condition depends heavily on the parameter χ: if we consider a different parameter $\chi' \in (0, \chi)$, then it is not clear how the respective matrices A and B defined in terms of this new parameter are related (in particular, we might not have an overlap). On the other hand, the estimate of the second fraction is entirely dynamical, and does not depend on the comparison of A and B. Because of this, if we consider a real orbit as in Corollary 4.12, then the functions $S_{\chi'}, U_{\chi'}$ defined in terms of χ' satisfy an improvement. This fact was observed in [**BO21**]. Let us recall the context of Corollary 4.12. For a fixed $\chi' \in (0, \chi)$, let $\delta = (e^{-\chi'} - e^{-\chi})^{3/\beta} < 1$. Fix $\widehat{x} \in \text{NUH}^*$, and let $\widehat{x}_n = \widehat{f}^n(\widehat{x})$, $p_n^s = q^s(\widehat{x}_n)$, $p_n^u = q^u(\widehat{x}_n)$. The sequence $\underline{v} = \{\Psi_{\widehat{x}_n}^{\delta p_n^s, \delta p_n^u}\}_{n\in\mathbb{Z}}$ defines sets $\widehat{V}_{\chi'}^s[\underline{v}]$ and $\widehat{V}_{\chi'}^u[\underline{v}]$. Clearly, if $\widehat{y} \in \widehat{V}_{\chi'}^s[\underline{v}]$ then $\widehat{f}(\widehat{y}) \in \widehat{f}(\widehat{V}_{\chi'}^s[\underline{v}]) \subset \widehat{V}_{\chi'}^s[\sigma(\underline{v})]$.

Corollary 6.5. *The following holds for all $\varepsilon > 0$ small enough. For $\widehat{y} \in \widehat{V}_{\chi'}^s[\underline{v}]$, let $v_1 \in T_{\widehat{f}(\widehat{y})}\widehat{V}_{\chi'}^s[\sigma(\underline{v})]$ and $v_0 = \widehat{df}^{(-1)} v_1 \in T_{\widehat{y}}\widehat{V}_{\chi'}^s[\underline{v}]$. Also, let $w_0 \in \widehat{TM}_{\widehat{x}}$ s.t. $v_0 = \Theta_{\widehat{x},\widehat{y}}^s(w_0)$ and $w_1 \in \widehat{TM}_{\widehat{f}(\widehat{x})}$ s.t. $v_1 = \Theta_{\widehat{f}(\widehat{x}),\widehat{f}(\widehat{y})}^s(w_1)$. If $\frac{S_{\chi'}(\widehat{f}(\widehat{y}),v_1)}{S_{\chi'}(\widehat{f}(\widehat{x}),w_1)} = \exp[\pm\xi]$ for $\xi \geq \sqrt{\varepsilon}$, then*

$$\frac{S_{\chi'}(\widehat{y}, v_0)}{S_{\chi'}(\widehat{x}, w_0)} = \exp\left[\pm(\xi - Q(\widehat{x})^{\beta/4})\right].$$

An analogous statement holds for unstable sets.

Proof. As in Lemma 6.4, it is enough to prove the result for $\widehat{V}_{\chi'}^s[\underline{v}]$. We assume that $\|v_0\| = 1$ and write $S_{\chi'}(\cdot, v)$ simply by $\|\|v\|\|_{\chi'}$. Again, we focus on one side of the estimate: assuming that $\frac{\|\|v_1\|\|_{\chi'}}{\|\|w_1\|\|_{\chi'}} \leq \exp[\xi]$ for $\xi \geq \sqrt{\varepsilon}$, we will prove that $\frac{|||v_0|||_{\chi'}}{|||w_0|||_{\chi'}} \le \exp[\xi - Q(\widehat{x})^{\beta/4}]$. Note that, in the notation of Lemma 6.4, we have $w_0 = \varrho$, and so we can obtain as in the estimate of $\frac{|||v_0|||}{|||\varrho|||}$ in Lemma 6.4 that

$$\begin{aligned}\frac{|||v_0|||^2_{\chi'}}{|||w_0|||^2_{\chi'}} &= \frac{2\|v_0\|^2 + e^{2\chi'}|||v_1|||^2_{\chi'}}{2\|w_0\|^2 + e^{2\chi'}|||w_1|||^2_{\chi'}} \\ &\le \frac{2\|v_0\|^2 + e^{2\chi'}e^{2\xi}|||w_1|||^2_{\chi'}}{2\|w_0\|^2 + e^{2\chi'}|||w_1|||^2_{\chi'}} \\ &= e^{2\xi}\left(1 - 2\frac{\|w_0\|^2 - \frac{\|v_0\|^2}{e^{2\xi}}}{|||w_0|||^2_{\chi'}}\right).\end{aligned}$$

Therefore, we just need to prove that

$$\frac{\|w_0\|^2 - e^{-2\xi}}{|||w_0|||^2_{\chi'}} \ge Q(\widehat{x})^{\beta/4} \iff \|w_0\|^2 - e^{-2\xi} \ge |||w_0|||^2_{\chi'} Q(\widehat{x})^{\beta/4}.$$

Note that:

- $|\|w_0\| - 1| < \varepsilon^{3/2}$ and so $\|w_0\|^2 - e^{-2\xi} > \varepsilon^{1/2} - 3\varepsilon^{3/2}$.
- Since $|||w_0|||_{\chi'} = \|C_{\chi'}(\widehat{x})^{-1}w_0\| \le \|C(\widehat{x})^{-1}w_0\| \le 2\|C(\widehat{x})^{-1}\|$, we have
$$|||w_0|||^2_{\chi'}Q(\widehat{x})^{\beta/4} \le 4\|C(\widehat{x})^{-1}\|^2 Q(\widehat{x})^{\beta/4} \le 4\varepsilon^{1/4}Q(\widehat{x})^{5\beta/24} \le 4\varepsilon^{3/2}.$$

If $\varepsilon > 0$ is small enough then $\varepsilon^{1/2} - 3\varepsilon^{3/2} > 4\varepsilon^{3/2}$, which concludes the proof. □

The above corollary indeed works for the (potentially larger) invariant sets associated to $\{\Psi^{p^s_n, p^u_n}_{\widehat{x}_n}\}_{n\in\mathbb{Z}}$, but since our application will require to diminish the Pesin charts by δ (so that Corollary 4.12 works), we preferred to state it with δ.

6.3. Proof that $\widehat{x} \in$ NUH

The first step in the proof of Theorem 6.1 is guaranteeing that $\widehat{x} \in$ NUH. We start proving that the relevance of each symbol of $\mathscr{A}$ implies the existence of stable/unstable sets where S/U are bounded. We continue employing the notation for tangent bundles of stable/unstable sets. Recall the definitions of $s(\widehat{x}), u(\widehat{x})$ on page 15. The result below is [**BO21**, Lemma 4.2 and Corollary 4.3], adapted to our context.

LEMMA 6.6. *Let $\widehat{W}^s = \widehat{V}^s[\underline{w}^+]$, where $\underline{w}^+$ is a positive ε-gpo. If $\widehat{W}^s \cap \mathrm{NUH}^\# \neq \emptyset$ then*

$$\sup_{\widehat{y}\in\widehat{W}^s} s(\widehat{y}) < \infty.$$

The same applies to negative ε-gpo's, with respect to the function u.

PROOF. We start proving the lemma for stable sets. For $\chi' \in (0,\chi)$, represent $S_{\chi'}(\cdot, v)$ simply by $|||v|||_{\chi'}$ and $S(\cdot, v)$ by $|||v|||$.

CLAIM 1: For every v, $|||v||| = \sup_{\chi'<\chi}|||v|||_{\chi'}$. In particular, if $|||v|||_{\chi'} \le L$ for every $v \in T\widehat{W}^s$ with $\|v\| = 1$ and every $\chi' \in (0,\chi)$, then $\sup_{\widehat{y}\in\widehat{W}^s} s(\widehat{y}) \le L$.

The proof of Claim 1 is direct, and we leave the details to the reader. By hypothesis, there is $\widehat{x} \in \widehat{W}^s \cap \mathrm{NUH}^\#$. Write $\widehat{x}_n = \widehat{f}^n(\widehat{x})$, $p^s_n = q^s(\widehat{f}^n(\widehat{x}))$ and $p^u_n = q^u(\widehat{f}^n(\widehat{x}))$. We have $\limsup_{n\to\infty} q(\widehat{x}_n) > 0$ and so there is $q > 0$ and an increasing sequence $\{n_k\}_{k\geq 0}$ s.t. $q(\widehat{x}_{n_k}) \geq q$ for all $k \geq 0$. By (3.3) and Lemma 3.4, for every $\widehat{y} \in \mathrm{NUH}^*$ we have $\|C(\widehat{y})^{-1}\| < \varepsilon^{1/8} Q(\widehat{y})^{-\beta/48} < q(\widehat{y})^{-\beta/48}$ and so in particular $Q(\widehat{x}_{n_k}) > q$ and $\|C(\widehat{x}_{n_k})^{-1}\| < q^{-\beta/48}$ for every $k \geq 0$.

Fix $\chi' \in (0,\chi)$, and let $\overline{\chi} = \frac{\chi+\chi'}{2}$. Also, let $\underline{v} = \{v_n\}_{n\in\mathbb{Z}}$ with $v_n = \Psi_{\widehat{x}_n}^{\delta p_n^s, \delta p_n^u}$, where $\delta = (e^{-\overline{\chi}} - e^{-\chi})^{3/\beta} < 1$. Write $\widehat{V}^s_{\overline{\chi},n} = \widehat{V}^s_{\overline{\chi}}[\{v_\ell\}_{\ell\geq n}]$. We want to bound $\vert\vert\vert v\vert\vert\vert_{\chi'}$, uniformly in v with $\|v\| = 1$ and χ'. Instead of $\widehat{W}^s$, doing this for $\widehat{V}^s_{\overline{\chi},0}$ is simpler, since inside this latter set we have better estimates given by Corollary 4.12. Although $\widehat{W}^s$ is in general not contained in $\widehat{V}^s_{\overline{\chi},0}$, if n is large then $\widehat{f}^n(\widehat{W}^s) \subset \widehat{V}^s_{\overline{\chi},n}$. The reason for this inclusion is that, while by Proposition 4.11(1) the size of $\widehat{f}^n(\widehat{W}^s)$ is at most of the order of $e^{-\frac{\chi}{2}n}$, the size of $\widehat{V}^s_{\overline{\chi},n}$ is of the order of $q(\widehat{x}_n)$, which decreases to zero slower than $e^{-\varepsilon n}$. Let us provide the details of this fact.

Claim 2: If n is large enough then $\widehat{f}^n(\widehat{W}^s) \subset \widehat{V}^s_{\overline{\chi},n}$.

Proof of Claim 2. Write $\widehat{x} = \{x_n\}_{n\in\mathbb{Z}}$ and let W^s be the s-admissible manifold associated to $\widehat{W}^s$. We will check Proposition 4.7(3). Since $B^d[\delta p_n^s] \times B^{m-d}[\delta p_n^s] \supset B[\delta p_n^s]$, it is enough to prove that if n is large enough then $f^n(W^s) \subset \Psi_{\widehat{x}_n}(B[\delta p_n^s])$. By Proposition 4.7(4), $f^n(W^s) \subset B(x_n, e^{-\frac{\chi}{2}n})$. Since

$$\Psi_{\widehat{x}_n}(B[\delta p_n^s]) \supset B\left(x_n, \tfrac{1}{2}\|C(\widehat{x}_n)^{-1}\|^{-1}\delta p_n^s\right),$$

it is enough to prove that for n large enough

$$\tfrac{1}{2}\|C(\widehat{x}_n)^{-1}\|^{-1}\delta p_n^s > e^{-\frac{\chi}{2}n} \iff e^{-\frac{\chi}{2}n}\|C(\widehat{x}_n)^{-1}\|\tfrac{1}{p_n^s} < \tfrac{\delta}{2}.$$

Since $\|C(\widehat{x}_n)^{-1}\|(p_n^s)^{\beta/48} \leq \|C(\widehat{x}_n)^{-1}\| Q(\widehat{x}_n)^{\beta/48} < 1$ and $p_n^s \geq e^{-\varepsilon n} p_0^s$, we have

$$\begin{aligned} e^{-\frac{\chi}{2}n}\|C(\widehat{x}_n)^{-1}\|\tfrac{1}{p_n^s} &< e^{-\frac{\chi}{2}n}\left(\tfrac{1}{p_n^s}\right)^{1+\frac{\beta}{48}} \leq e^{-\frac{\chi}{2}n}\left(\tfrac{e^{\varepsilon n}}{p_0^s}\right)^{1+\frac{\beta}{48}} \\ &= \left(\tfrac{1}{p_0^s}\right)^{1+\frac{\beta}{48}} e^{-[\frac{\chi}{2} - \varepsilon(1+\frac{\beta}{48})]n} \end{aligned}$$

which, for $\varepsilon > 0$ small enough, converges to zero exponentially fast. $\square$

Now we complete the proof of the lemma for $\widehat{W}^s$. Start observing that, for every $n \geq 0$, Corollary 4.12 implies that if $v \in T\widehat{V}^s_{\overline{\chi},n}$ with $\|v\| = 1$ then

$$\begin{aligned} \vert\vert\vert v\vert\vert\vert_{\chi'}^2 &= 2\sum_{\ell\geq 0} e^{2\ell\chi'}\|\widehat{df}^{(\ell)} v\|^2 \\ &\leq 2\sum_{\ell\geq 0} e^{2\ell\chi'} 64 e^{-2\ell\overline{\chi}}\|C(\widehat{x}_n)^{-1}\|^2 \\ &\leq 128\|C(\widehat{x}_n)^{-1}\|^2 \sum_{\ell\geq 0} e^{\ell(\chi'-\chi)} \end{aligned}$$

and so for every $k \geq 0$ and every $v \in T\widehat{V}^s_{\overline{\chi},n_k}$ we obtain the bound

$$\vert\vert\vert v\vert\vert\vert_{\chi'}^2 \leq 128 q^{-\beta/24}\sum_{\ell\geq 0} e^{\ell(\chi'-\chi)}.$$

Call this bound L^2, so that $\vert\vert\vert v\vert\vert\vert_{\chi'} \leq L$ for every $v \in T\widehat{V}^s_{\overline{\chi},n_k}$ with $\|v\| = 1$. Had L been independent of χ', the proof would be complete, but unfortunately $L \to \infty$ as $\chi' \to \chi$. The way to improve the estimate on $\vert\vert\vert v\vert\vert\vert_{\chi'}$ to a bound that is independent of χ' is by applying Corollary 6.5, as we now explain. Define $\xi \geq \sqrt{\varepsilon}$ by $e^{\xi} = \max\{\sqrt{2}L, e^{\sqrt{\varepsilon}}\}$. We claim that

$$\frac{\left|\left|\left|\Theta^s_{\widehat{x}_{n_k},\widehat{y}}(v)\right|\right|\right|_{\chi'}}{|||v|||_{\chi'}} = \exp[\pm\xi], \quad \text{for all } \widehat{y} \in \widehat{V}^s_{\overline{\chi},n_k} \text{ and } v \in E^s_{\widehat{x}_{n_k}}\backslash\{0\}.$$

By a normalization, we just need to check this estimate for $\|v\| = 1$. By Lemma 6.3(1), we have $\frac{1}{2} \leq \|\Theta^s_{\widehat{x}_{n_k},\widehat{y}}(v)\| \leq 2$ and so

$$(\sqrt{2}L)^{-1} = \frac{\sqrt{2}/2}{L} \leq \frac{\left|\left|\left|\Theta^s_{\widehat{x}_{n_k},\widehat{y}}(v)\right|\right|\right|_{\chi'}}{|||v|||_{\chi'}} \leq \frac{2L}{\sqrt{2}} = \sqrt{2}L.$$

Now fix $k \geq 1$. Apply Corollary 6.5 along the path $\widehat{x}_{n_{k-1}} \to \cdots \to \widehat{x}_{n_k}$. Since the ratio does not get worse for all transitions $\widehat{x}_\ell \to \widehat{x}_{\ell+1}$ and it improves a fixed amount in the last edge $\widehat{x}_{n_{k-1}} \to \widehat{x}_{n_{k-1}+1}$, we conclude that

$$\frac{\left|\left|\left|\Theta^s_{\widehat{x}_{n_{k-1}},\widehat{y}}(v)\right|\right|\right|_{\chi'}}{|||v|||_{\chi'}} = \exp\left[\pm(\xi - q^{\beta/4})\right], \quad \text{for all } \widehat{y} \in \widehat{V}^s_{\overline{\chi},n_{k-1}} \text{ and } v \in E^s_{\widehat{x}_{n_{k-1}}}\backslash\{0\}.$$

Repeating this procedure until reaching $\widehat{x}_{n_0}$, we obtain at least k improvements, as long as the ratio remains outside $[\exp(-\sqrt{\varepsilon}), \exp(\sqrt{\varepsilon})]$. Taking $k \to +\infty$, we conclude that

$$\frac{\left|\left|\left|\Theta^s_{\widehat{x}_{n_0},\widehat{y}}(v)\right|\right|\right|_{\chi'}}{|||v|||_{\chi'}} = \exp\left[\pm\sqrt{\varepsilon}\right]$$

for all $\widehat{y} \in \widehat{V}^s_{\overline{\chi},n_0}$ and $v \in E^s_{\widehat{x}_{n_0}}\backslash\{0\}$. Similarly, we obtain that for every $k \geq 0$ it holds

$$\frac{\left|\left|\left|\Theta^s_{\widehat{x}_{n_k},\widehat{y}}(v)\right|\right|\right|_{\chi'}}{|||v|||_{\chi'}} = \exp[\pm\sqrt{\varepsilon}], \quad \text{for all } \widehat{y} \in \widehat{V}^s_{\overline{\chi},n_k} \text{ and } v \in E^s_{\widehat{x}_{n_k}}\backslash\{0\}. \tag{6.4}$$

Now fix k large enough s.t. $\widehat{f}^{n_k}(\widehat{W}^s) \subset \widehat{V}^s_{\overline{\chi},n_k}$. Let $v \in T\widehat{W}^s$ with $\|v\| = 1$. If $w = \widehat{df}^{(n_k)} v \in T\widehat{V}^s_{\overline{\chi},n_k}$ then

$$\begin{aligned}
|||v|||^2_{\chi'} &= 2\sum_{n\geq 0} e^{2n\chi'}\|\widehat{df}^{(n)} v\|^2 \\
&= 2\sum_{n=0}^{n_k-1} e^{2n\chi'}\|\widehat{df}^{(n)} v\|^2 + 2\sum_{n\geq n_k} e^{2n\chi'}\|\widehat{df}^{(n)} v\|^2 \\
&= 2\sum_{n=0}^{n_k-1} e^{2n\chi'}\|\widehat{df}^{(n)} v\|^2 + e^{2n_k\chi'}\|w\|^2 \cdot \left|\left|\left|\tfrac{w}{\|w\|}\right|\right|\right|^2_{\chi'} \\
&\leq 2\sum_{n=0}^{n_k-1} e^{2n\chi}\|\widehat{df}^{(n)} v\|^2 + e^{2n_k\chi}\|w\|^2 \cdot \left|\left|\left|\tfrac{w}{\|w\|}\right|\right|\right|^2_{\chi'}.
\end{aligned}$$

If $\widehat{z}$ is the center of the zeroth chart defining $\widehat{W}^s = \widehat{V}^s[\underline{w}^+]$, then by Proposition 4.11(1) the first term above is bounded by $128\|C(\widehat{z})^{-1}\|^2 \sum_{n=0}^{n_k-1} e^{n\chi}$. To estimate the second term, write $w \in T_{\widehat{y}}\widehat{V}^s_{\overline{\chi},n_k}$ and define ϱ by the equality $\frac{w}{\|w\|} = \Theta^s_{\widehat{x}_{n_k},\widehat{y}}(\varrho)$.

Using estimate (6.4), Lemma 6.3(1) and Lemma 3.1(1), we get that

$$\begin{aligned}\left|\left|\left|\tfrac{w}{\|w\|}\right|\right|\right|_{\chi'} &\le e^{\sqrt{\varepsilon}}|||\varrho|||_{\chi'} \le 2e^{\sqrt{\varepsilon}}\left|\left|\left|\tfrac{\varrho}{\|\varrho\|}\right|\right|\right|_{\chi'}\\ &\le 2e^{\sqrt{\varepsilon}}s(\widehat{x}_{n_k}) \le 2e^{\sqrt{\varepsilon}}\|C(\widehat{x}_{n_k})^{-1}\| < 2e^{\sqrt{\varepsilon}}q^{-\beta/48}.\end{aligned}$$

Hence $e^{2n_k\chi}\|w\|^2 \cdot \left|\left|\left|\tfrac{w}{\|w\|}\right|\right|\right|_{\chi'}^2 \le 256\|C(\widehat{z})^{-1}\|^2 e^{n_k\chi+2\sqrt{\varepsilon}}q^{-\beta/24}$, and so

$$|||v|||_{\chi'}^2 \le 128\|C(\widehat{z})^{-1}\|^2\left[\sum_{n=0}^{n_k-1}e^{n\chi} + 2e^{n_k\chi+2\sqrt{\varepsilon}}q^{-\beta/24}\right],$$

which is independent of v and χ'. By Claim 1, the proof for $\widehat{W}^s$ is complete.

Now let $\widehat{W}^u = \widehat{V}^u[\underline{w}^-]$ for a negative ε-gpo $\underline{w}^- = \{\Psi_{\widehat{y}_n}^{q_n^s,q_n^u}\}_{n\le 0}$, and assume that there exists $\widehat{x} \in \widehat{W}^u \cap \mathrm{NUH}^\#$. As before, write $\widehat{x}_n = \widehat{f}^n(\widehat{x})$ and let $\widehat{V}^u_{\overline{\chi},n} = \widehat{V}^u_{\overline{\chi}}[\{\Psi_{\widehat{x}_k}^{\delta p_k^s,\delta p_k^u}\}_{k\le n}]$ for $n \le 0$. The proof for $\widehat{W}^u$ is identical to the proof for $\widehat{W}^s$, provided the inverse branches used to define $\widehat{W}^u$ are the same as those used to define $\widehat{V}^u_{\overline{\chi},n}$. In other words, we just need to verify that $f_{\widehat{x}_n}^{-1}$ and $f_{\widehat{y}_n}^{-1}$ are the same inverse branch for all $n < 0$. Fix $n < 0$. By Lemma 4.8(2), $x_{n+1} \in \Psi_{\widehat{y}_{n+1}}(B[20Q(\widehat{y}_{n+1})])$. Since $f_{\widehat{x}_n}^{-1}(x_{n+1}) = x_n = f_{\widehat{y}_n}^{-1}(x_{n+1})$, the uniqueness of inverse branches provided by assumption (A5) implies that $f_{\widehat{x}_n}^{-1}$ and $f_{\widehat{y}_n}^{-1}$ are the same inverse branches. □

Now we proceed to show that, in the notation of Theorem 6.1, $\pi(\underline{v}) \in \mathrm{NUH}$. Recall the relevance property of each ε-double chart of $\mathscr{A}$, see Theorem 5.1(3).

PROPOSITION 6.7. *For every $v_0 \in \mathscr{A}$, there exists a constant $L = L(v_0)$ s.t. the following holds. If $\underline{v} = \{v_n\}_{n\in\mathbb{Z}} \in \Sigma^\#$ satisfies $v_n = v_0$ for infinitely many $n > 0$ and if $\widehat{x} = \pi(\underline{v})$, then $s(\widehat{x}) < L$. The same applies to $u(\widehat{x})$. In particular, $\widehat{x} \in \mathrm{NUH}$.*

PROOF. As usual, we represent $S(\cdot, v)$ simply by $|||v|||$. Write $\widehat{V}^s = \widehat{V}^s[\underline{v}]$, $v_n = \Psi_{\widehat{x}_n}^{p_n^s,p_n^u}$, $\eta_n = p_n^s \wedge p_n^u$, and let $\{n_k\}_{k\ge1}$ be an increasing sequence s.t. $v_{n_k} = v_0$. Since v_0 is relevant, there is $\underline{w} \in \Sigma$ with $w_0 = v_0$ s.t. $\pi(\underline{w}) \in \mathrm{NUH}^\#$. By Lemma 6.6, if $\widehat{W}^s = \widehat{V}^s[\underline{w}]$ then $\sup_{\widehat{y}\in\widehat{W}^s} s(\widehat{y}) < \infty$. By Lemma 6.3(1),

$$L_0 := \sup\left\{\frac{|||\Theta^s_{\widehat{x}_0,\widehat{y}}(v)|||}{|||v|||} : \begin{array}{c}\widehat{y} \in \widehat{W}^s, \vartheta[\widehat{y}] \in \Psi_{\widehat{x}_0}(B^d[\eta_0] \times B^{m-d}[\eta_0]),\\ v \in E^s_{\widehat{x}_0}\setminus\{0\}\end{array}\right\}$$

is also finite. Define $L_1 = \max\{L_0, e^{\sqrt{\varepsilon}}\} > 1$.

For each $k \ge 1$, we have $\widehat{W}^s \in \widehat{\mathscr{M}}^s(v_{n_k})$. Starting from $\widehat{W}^s$, apply the stable graph transform along the path $v_0 \overset{\varepsilon}{\to} v_1 \overset{\varepsilon}{\to} \cdots \overset{\varepsilon}{\to} v_{n_k}$ to obtain some stable set at v_0, call it $\widehat{W}^s_k$. Let G be the representing function of $\widehat{V}^s$, and let G_k be the representing function of $\widehat{W}^s_k$. By Proposition 4.7(1), $\|G_k - G\|_{C^1} \xrightarrow[k\to\infty]{} 0$.

CLAIM: If $\{w_k\}_{k\ge1} \subset TM$ converges to $w \in TM$ in the Sasaki metric, then

$$\sum_{n\ge0} e^{2n\chi}\|\widehat{df}^{(n)}w\|^2 \le \liminf_{k\to+\infty}\sum_{n\ge0} e^{2n\chi}\|\widehat{df}^{(n)}w_k\|^2.$$

Proof of the claim. Fix $N > 0$. For each $\delta > 0$, there exists $k_0 \geq 1$ s.t.

$$k \geq k_0 \Longrightarrow \frac{\|\widehat{df}^{(n)} w_k\|}{\|\widehat{df}^{(n)} w\|} = e^{\pm\delta} \text{ for } n = 0, 1, \ldots, N.$$

For $k \geq k_0$, we have

$$\sum_{n=0}^{N} e^{2n\chi} \|\widehat{df}^{(n)} w\|^2 \leq e^{2\delta} \sum_{n=0}^{N} e^{2n\chi} \|\widehat{df}^{(n)} w_k\|^2 \leq e^{2\delta} \sum_{n\geq 0} e^{2n\chi} \|\widehat{df}^{(n)} w_k\|^2$$

and so

$$\sum_{n=0}^{N} e^{2n\chi} \|\widehat{df}^{(n)} w\|^2 \leq e^{2\delta} \liminf_{k\to+\infty} \sum_{n\geq 0} e^{2n\chi} \|\widehat{df}^{(n)} w_k\|^2.$$

Since $\delta > 0$ is arbitrary, we get that

$$\sum_{n=0}^{N} e^{2n\chi} \|\widehat{df}^{(n)} w\|^2 \leq \liminf_{k\to+\infty} \sum_{n\geq 0} e^{2n\chi} \|\widehat{df}^{(n)} w_k\|^2.$$

Finally, take $N \to +\infty$ to conclude the proof of the claim. □

Write $\vartheta[\widehat{x}] = \Psi_{\widehat{x}_0}(z, G(z))$. Since $\widehat{x} = \pi(\underline{v})$, by Lemma 4.4 we have that $\|z\| < \eta_0$. For each $k \geq 1$, define $\widehat{y}_k$ as the unique element of $\widehat{W}^s_k$ s.t. $\vartheta[\widehat{y}_k] = \Psi_{\widehat{x}_0}(z, G_k(z))$. Fix $v \in E^s_{\widehat{x}_0}$ with $\|v\| = 1$. If $v = (d\Psi_{\widehat{x}_0})_0 \begin{bmatrix} w \\ 0 \end{bmatrix}$ then

$$\Theta^s_{\widehat{x}_0,\widehat{x}}(v) = (d\Psi_{\widehat{x}_0})_{(z,G(z))} \begin{bmatrix} w \\ (dG)_z w \end{bmatrix} \quad \text{and} \quad \Theta^s_{\widehat{x}_0,\widehat{y}_k}(v) = (d\Psi_{\widehat{x}_0})_{(z,G_k(z))} \begin{bmatrix} w \\ (dG_k)_z w \end{bmatrix}$$

and so $\Theta^s_{\widehat{x}_0,\widehat{y}_k}(v) \to \Theta^s_{\widehat{x}_0,\widehat{x}}(v)$ in the Sasaki metric. By the claim,

$$\left|\left|\left|\Theta^s_{\widehat{x}_0,\widehat{x}}(v)\right|\right|\right| \leq \liminf_{k\to+\infty} \left|\left|\left|\Theta^s_{\widehat{x}_0,\widehat{y}_k}(v)\right|\right|\right|$$

and so it is enough to bound the right hand side above.

We claim that $\vartheta[\widehat{f}^n(\widehat{y}_k)] \in \Psi_{\widehat{x}_n}(B^d[\eta_n] \times B^{m-d}(\eta_n))$ for $n = 0, \ldots, n_k$. To prove this, note that $\widehat{f}^n(\widehat{y}_k)$ belongs to a stable set and so it is enough to show that, in the charts representation, the first coordinate of $\vartheta[\widehat{f}^n(\widehat{y}_k)]$ belongs to $B^d[\eta_n]$. The case $n = 0$ is true because $\|z\| < \eta_0$. The proof is by induction, so we just show how to obtain it for $n = 1$. Write $F_{\widehat{x}_0,\widehat{x}_1} = D(\widehat{x}_0) + (H_1^+, H_2^+)$, where $D(\widehat{x}_0)$ is given by Lemma 3.1(2) and $H^+ = (H_1^+, H_2^+)$ satisfies itens (a)–(c) in Theorem 4.3. We have $\vartheta[\widehat{y}_k] = \Psi_{\widehat{x}_0}(z, G_k(z))$ and so $\vartheta[\widehat{f}(\widehat{y}_k)] = \Psi_{\widehat{x}_1}(\overline{z}, *)$ where $\overline{z} = D_s(\widehat{x}_0)z + H_1^+(z, G_k(z))$. Then

$$\begin{aligned}
\|\overline{z}\| &\leq \|D_s(\widehat{x}_0)z\| + \left\|H_1^+(z, G_k(z))\right\| \\
&\leq \|D_s(\widehat{x}_0)z\| + \left\|H^+(z, G_k(z))\right\| \\
&\leq \|D_s(\widehat{x}_0)\|\|z\| + \left\|H^+(0,0)\right\| + \left\|dH^+\right\|_{C^0(B[2\eta_0])} \|(z, G_k(z))\| \\
&\leq e^{-\chi}\eta_0 + \varepsilon\eta_0 + \varepsilon(2\eta_0)^{\beta/3} 2\eta_0 \\
&\leq (e^{-\chi} + 3\varepsilon)\eta_0 \\
&\leq (e^{-\chi} + 3\varepsilon)e^{\varepsilon}\eta_1
\end{aligned}$$

is smaller than η_1 for $\varepsilon > 0$ small enough. Now, for fixed $k \geq 1$ write $\varrho_k = \Theta^s_{\widehat{x}_0,\widehat{y}_k}(v)$

and define $w_\ell \in E^s_{\widehat{x}_\ell}$ by the equality $\Theta^s_{\widehat{x}_\ell,\widehat{f}^\ell(\widehat{y}_k)}(w_\ell) = \widehat{df}^{(\ell)}\varrho_k$, for all $\ell \geq 0$. Since $\widehat{f}^{n_k}(\widehat{y}_k) \in \widehat{f}^{n_k}(\widehat{W}^s_k) \subset \widehat{W}^s$, we have that

$$\frac{\left\|\left\|\widehat{df}^{n_k}\varrho_k\right\|\right\|}{\|\|w_{n_k}\|\|} \leq L_1.$$

By Lemma 6.4, we get that

$$\frac{\left\|\left\|\widehat{df}^{n_k-1}\varrho_k\right\|\right\|}{\|\|w_{n_k-1}\|\|} \leq L_1$$

and, repeating this procedure, we obtain that $\frac{\|\|\varrho_k\|\|}{\|\|w_0\|\|} \leq L_1$. By Lemma 6.3(1), $\|\|\varrho_k\|\| \leq L_1\|\|w_0\|\| \leq 2L_1 s(\widehat{x}_0)$ and so

$$\left\|\left\|\Theta^s_{\widehat{x}_0,\widehat{x}}(v)\right\|\right\| \leq 2L_1 s(\widehat{x}_0). \tag{6.5}$$

Defining $L = L(v_0) := 3L_1 s(\widehat{x}_0)$ and applying Lemma 6.3(1) again, it follows that $\|\|v\|\| < L$ for all $v \in T_{\widehat{x}}\widehat{V}^s$ with $\|v\| = 1$, and so $s(\widehat{x}) < L$.

The proof for $\widehat{V}^u[\underline{v}]$ is identical to the above since, as in the proof of Lemma 6.6, the inverse branches of the unstable sets involved in the proof coincide. Finally, we show that $\widehat{x} \in$ NUH. Defining $E^s_{\widehat{x}} = T_{\widehat{x}}(\widehat{V}^s[\underline{v}])$ and $E^u_{\widehat{x}} = T_{\widehat{x}}(\widehat{V}^u[\underline{v}])$, then what we just proved and Proposition 4.11(2) imply that $\widehat{x}$ satisfies (NUH1)–(NUH3). □

Consequently, if $\underline{v} \in \Sigma^\#$ then $\widehat{x} = \pi(\underline{v}) \in$ NUH and so the Lyapunov inner product $\|\|\cdot\|\|$ is well-defined in $\widehat{TM}_{\widehat{x}}$, as well as $C(\widehat{x}), Q(\widehat{x})$. In the next sections we will prove parts (1)–(5) of Theorem 6.1.

6.4. Control of d and ρ

We start estimating $d(\vartheta[\widehat{x}_0], \vartheta[\widehat{x}])$. By Lemma 4.4, $\vartheta[\widehat{x}] = \Psi_{\widehat{x}_0}(w)$ for some $\|w\| < 50^{-1}(p^s_0 \wedge p^u_0)$. Since $\vartheta[\widehat{x}_0] = \Psi_{\widehat{x}_0}(0)$ and $\Psi_{\widehat{x}_0}$ is 2-Lipschitz, we get that $d(\vartheta[\widehat{x}_0], \vartheta[\widehat{x}]) < 25^{-1}(p^s_0 \wedge p^u_0)$. Now we estimate $\frac{\rho(\widehat{x}_0)}{\rho(\widehat{x})}$. Such estimate *does not* follow from Proposition 4.1(1), since we assume no overlap condition between $\widehat{x}_0$ and $\widehat{x}$. Nevertheless, we can employ a calculation similar to that used to prove Proposition 4.1(1). Start observing that, by the above estimate, $d(\vartheta[\widehat{x}_0], \vartheta[\widehat{x}]) < 25^{-1}(p^s_0 \wedge p^u_0) < \varepsilon^2\rho(\widehat{x}_0)^2 < \varepsilon^2 d(\vartheta[\widehat{x}_0], \mathscr{S})$ and so

$$d(\vartheta[\widehat{x}], \mathscr{S}) = d(\vartheta[\widehat{x}_0], \mathscr{S}) \pm d(\vartheta[\widehat{x}_0], \vartheta[\widehat{x}]) = (1 \pm \varepsilon^2) d(\vartheta[\widehat{x}_0], \mathscr{S}) = e^{\pm\varepsilon} d(\vartheta[\widehat{x}_0], \mathscr{S}).$$

To estimate $\frac{d(\vartheta_1[\widehat{x}],\mathscr{S})}{d(\vartheta_1[\widehat{x}_0],\mathscr{S})}$, start observing that, since $\vartheta[\widehat{x}] \in D_{\widehat{x}_0}$, assumption (A6) implies that $d(\vartheta_1[\widehat{x}], \vartheta_1[\widehat{x}_0]) < \varepsilon^2\rho(\widehat{x}_0) < \varepsilon^2 d(\vartheta_1[\widehat{x}_0], \mathscr{S})$ and so

$$\begin{aligned} d(\vartheta_1[\widehat{x}], \mathscr{S}) &= d(\vartheta_1[\widehat{x}_0], \mathscr{S}) \pm d(\vartheta_1[\widehat{x}_0], \vartheta_1[\widehat{x}]) \\ &= (1 \pm \varepsilon^2) d(\vartheta_1[\widehat{x}_0], \mathscr{S}) \\ &= e^{\pm\varepsilon} d(\vartheta_1[\widehat{x}_0], \mathscr{S}). \end{aligned}$$

We estimate $\frac{d(\vartheta_{-1}[\widehat{x}],\mathscr{S})}{d(\vartheta_{-1}[\widehat{x}_0],\mathscr{S})}$ in the same way, using that $\vartheta[\widehat{x}] \in E_{\widehat{f}^{-1}(\widehat{x}_0)}$. Altogether, we conclude that $\frac{\rho(\widehat{x}_0)}{\rho(\widehat{x})} = e^{\pm\varepsilon}$. This proves Theorem 6.1 part (1).

6.5. Control of C^{-1}

Recall that $\widehat{x} = \pi(\underline{v})$ for $\underline{v} \in \Sigma^{\#}$. We proved in Section 6.3 that $\widehat{x} \in$ NUH, i.e., there is a splitting $\widehat{TM}_{\widehat{x}} = E^s_{\widehat{x}} \oplus E^u_{\widehat{x}}$ satisfying (NUH1)–(NUH3). To control C^{-1}, we need to control the Lyapunov inner product for vectors in $E^s_{\widehat{x}}$ and $E^u_{\widehat{x}}$. We explain how to make the control in $E^s_{\widehat{x}}$ (the control in $E^u_{\widehat{x}}$ is analogous). Write $\underline{v} = \{v_n\}_{n\in\mathbb{Z}}$ and $\Theta_n = \Theta_{\widehat{x}_n, \widehat{f}^n(\widehat{x})}$. Without loss of generality, assume that v_0 repeats infinitely often in the future, i.e. there is an increasing sequence $\{n_k\}_{k\geq 1}$ s.t. $v_{n_k} = v_0$ for all $k \geq 1$. Let $L = 3L_1 s(\widehat{x}_0)$ as in Proposition 6.7, and let $\xi > 0$ s.t. $L = e^{\xi}$. Since $L_1 \geq e^{\sqrt{\varepsilon}}$ and $s(\widehat{x}_0) \geq \sqrt{2}$, we have $\xi > \sqrt{\varepsilon}$. We claim that

$$\frac{|||v|||}{|||\Theta_{n_k}(v)|||} = \exp[\pm\xi], \quad \text{for all } v \in E^s_{\widehat{x}_0}\backslash\{0\}. \tag{6.6}$$

By a normalization, we just need to check this for $\|v\| = 1$. We proved in Proposition 6.7 that $|||\Theta_{n_k}(v)||| \leq 2L_1 s(\widehat{x}_0)$, see estimate (6.5). On one hand, applying Lemma 6.3(1) we have $\frac{|||v|||}{|||\Theta_{n_k}(v)|||} \leq 2s(\widehat{x}_0) < L$, and on the other hand $\frac{|||v|||}{|||\Theta_{n_k}(v)|||} \geq \frac{\sqrt{2}}{2L_1 s(\widehat{x}_0)} > L^{-1}$, which proves (6.6). Now fix $k \geq 1$. Using (6.6), apply Lemma 6.4 along the path $v_{n_{k-1}} \overset{\varepsilon}{\to} \cdots \overset{\varepsilon}{\to} v_{n_k}$. Since the ratio does not get worse for all edges $v_\ell \overset{\varepsilon}{\to} v_{\ell+1}$ and it improves a fixed amount in the last edge $v_{n_{k-1}} \overset{\varepsilon}{\to} v_{n_{k-1}+1}$, we conclude that $\frac{|||v|||}{|||\Theta_{n_{k-1}}(v)|||} = \exp\left[\pm(\xi - Q(\widehat{x}_0)^{\beta/4})\right]$ for all $v \in E^s_{\widehat{x}_0}\backslash\{0\}$. Repeating this procedure until reaching v_0, we obtain at least k improvements, as long as the ratio remains outside $[\exp(-\sqrt{\varepsilon}), \exp(\sqrt{\varepsilon})]$. Taking $k \to +\infty$, we conclude that $\frac{|||v|||}{|||\Theta_0(v)|||} = \exp[\pm\sqrt{\varepsilon}]$ for all $v \in E^s_{\widehat{x}_0}\backslash\{0\}$. Since $v_{n_k} = v_0$, we obtain similarly that $\frac{|||v|||}{|||\Theta_{n_k}(v)|||} = \exp[\pm\sqrt{\varepsilon}]$ for all $v \in E^s_{\widehat{x}_0}\backslash\{0\}$. Finally, given $n \in \mathbb{Z}$, let $n_k > n$ and apply Lemma 6.4 along the path $v_n \overset{\varepsilon}{\to} \cdots \overset{\varepsilon}{\to} v_{n_k}$ to conclude that

$$\frac{|||v|||}{|||\Theta_n(v)|||} = \exp[\pm\sqrt{\varepsilon}], \quad \text{for all } v \in E^s_{\widehat{x}_n}\backslash\{0\}. \tag{6.7}$$

By a similar argument, we conclude that

$$\frac{|||v|||}{|||\Theta_n(v)|||} = \exp[\pm\sqrt{\varepsilon}], \quad \text{for all } v \in E^u_{\widehat{x}_n}\backslash\{0\}. \tag{6.8}$$

We are now ready to prove part (2) of Theorem 6.1. For simplicity, assume $n = 0$ and write $\Theta = \Theta_0$. If $v = v^s + v^u \in E^s_{\widehat{x}_0} \oplus E^u_{\widehat{x}_0}$, then $\Theta(v) = \Theta(v^s) + \Theta(v^u) \in E^s_{\widehat{x}} \oplus E^u_{\widehat{x}}$. By the calculation made in the proof of Lemma 3.1(1) and estimates (6.7) and (6.8),

$$\frac{\left\|C(\widehat{x}_0)^{-1}v\right\|^2}{\left\|C(\widehat{x})^{-1}\Theta(v)\right\|^2} = \frac{|||v^s|||^2 + |||v^u|||^2}{|||\Theta(v^s)|||^2 + |||\Theta(v^u)|||^2} = \exp[\pm 2\sqrt{\varepsilon}] \tag{6.9}$$

and so $\frac{\|C(\widehat{x}_0)^{-1}\|}{\|C(\widehat{x})^{-1}\Theta\|} = \exp[\pm\sqrt{\varepsilon}]$. By Lemma 6.3(2), if $\varepsilon > 0$ is small enough then $\left\|\Theta^{\pm 1}\right\| = \exp[\pm\sqrt{\varepsilon}]$, hence $\frac{\|C(\widehat{x}_0)^{-1}\|}{\|C(\widehat{x})^{-1}\|} = \exp[\pm 2\sqrt{\varepsilon}]$.

6.6. Control of Q, p^s, p^u and proof that $\widehat{x} \in \mathrm{NUH}^\#$

Now we prove parts (3) and (4) of Theorem 6.1. We begin controlling Q. As usual, let $n = 0$. Recall that $Q(\widehat{x}) = \max\{Q \in I_\varepsilon : Q \leq \widetilde{Q}(\widehat{x})\}$, where

$$\widetilde{Q}(\widehat{x}) = \varepsilon^{6/\beta} \min\left\{ \|C(\widehat{x})^{-1}\|^{-48/\beta}, \rho(\widehat{x})^{96a/\beta} \right\}.$$

By part (1), $\frac{\rho(\widehat{x}_0)^{96a/\beta}}{\rho(\widehat{x})^{96a/\beta}} = \exp\left[\pm\frac{96a\varepsilon}{\beta}\right] = \exp\left[\pm\frac{96\sqrt{\varepsilon}}{\beta}\right]$ for $\varepsilon > 0$ small enough. By part (2), $\frac{\|C(\widehat{x}_0)^{-1}\|^{-48/\beta}}{\|C(\widehat{x})^{-1}\|^{-48/\beta}} = \exp\left[\pm\frac{96\sqrt{\varepsilon}}{\beta}\right]$. Hence, $\frac{\widetilde{Q}(\widehat{x}_0)}{\widetilde{Q}(\widehat{x})} = \exp\left[\pm\frac{96\sqrt{\varepsilon}}{\beta}\right]$ and so $\frac{Q(\widehat{x}_0)}{Q(\widehat{x})} = \exp\left[\pm\left(\frac{96\sqrt{\varepsilon}}{\beta} + \frac{\varepsilon}{3}\right)\right]$. If $\varepsilon > 0$ is small enough then $\frac{Q(\widehat{x}_0)}{Q(\widehat{x})} = \exp[\pm\sqrt[3]{\varepsilon}]$.

Now we prove part (4). We show how to control p^s (the proof for p^u is analogous). By (GPO2), we have $p^s_0 = \delta_\varepsilon \inf\{e^{\varepsilon n} Q(\widehat{x}_n) : n \geq 0\}$ and so by part (3)

$$q^s(\widehat{x}) = \delta_\varepsilon \inf\{e^{\varepsilon n} Q(\widehat{f}^n(\widehat{x})) : n \geq 0\} = p^s_0 \times \exp\left[\pm\sqrt[3]{\varepsilon}\right].$$

Hence $q^s(\widehat{x}) > 0$ and $\frac{p^s_0}{q^s(\widehat{x})} = \exp[\pm\sqrt[3]{\varepsilon}]$.

Finally, observe that we also have $\frac{p^s_n \wedge p^u_n}{q(\widehat{f}^n(\widehat{x}))} = \exp[\pm\sqrt[3]{\varepsilon}]$. Since $\underline{v} \in \Sigma^\#$, it follows that $\limsup\limits_{n\to+\infty} q(\widehat{f}^n(\widehat{x})) > 0$ and $\limsup\limits_{n\to-\infty} q(\widehat{f}^n(\widehat{x})) > 0$, thus proving that $\widehat{x} \in \mathrm{NUH}^\#$.

6.7. Control of $\Psi^{-1}_{\widehat{f}^n(\widehat{x})} \circ \Psi_{\widehat{x}_n}$

Fix $n = 0$, and let $x_0 = \vartheta[\widehat{x}_0]$, $x = \vartheta[\widehat{x}]$, $\eta = p^s_0 \wedge p^u_0$, $\Theta = \Theta_{\widehat{x}_0,\widehat{x}}$, and $P = P_{x_0,x}$. We start proving that $\Psi^{-1}_{\widehat{x}} \circ \Psi_{\widehat{x}_0}$ is well-defined in $B[10Q(\widehat{x}_0)]$. We have

$$\Psi_{\widehat{x}_0}(B[10Q(\widehat{x}_0)]) \subset B(x_0, 20Q(\widehat{x}_0)) \subset B(x, 20Q(\widehat{x}_0) + d(x_0, x)) \subset \mathfrak{B}_{\widehat{x}},$$

where in the last inclusion we used parts (1) and (3) to obtain that $20Q(\widehat{x}_0) + d(x_0, x) < 20Q(\widehat{x}_0) + \frac{\eta}{25} < 25Q(\widehat{x}_0) \leq 25e^{\sqrt[3]{\varepsilon}} Q(\widehat{x}) < \mathfrak{d}(\widehat{x})$. By assumption (A1), $\Psi^{-1}_{\widehat{x}} \circ \Psi_{\widehat{x}_0}$ is well-defined in $B[10Q(\widehat{x}_0)]$.

Now we represent $\Psi^{-1}_{\widehat{x}} \circ \Psi_{\widehat{x}_0}$ as required. We have $\Psi^{-1}_{\widehat{x}} \circ \Psi_{\widehat{x}_0} = C(\widehat{x})^{-1} \circ \Phi \circ C(\widehat{x}_0)$, where $\Phi = \exp^{-1}_x \circ \exp_{x_0}$. The composition $C(\widehat{x})^{-1} \circ \Theta \circ C(\widehat{x}_0)$ has norm close to one. Indeed, by (6.9) we have $\|C(\widehat{x})^{-1} \circ \Theta \circ C(\widehat{x}_0) v\| = e^{\pm 2\sqrt{\varepsilon}} \|C(\widehat{x}_0)^{-1} \circ C(\widehat{x}_0) v\| = e^{\pm 2\sqrt{\varepsilon}} \|v\|$. By the polar decomposition for matrices, $C(\widehat{x})^{-1} \circ \Theta \circ C(\widehat{x}_0) = OR$ where O is an orthogonal matrix and R is positive symmetric with $\|Rv\| = e^{\pm 2\sqrt{\varepsilon}} \|v\|$ for all $v \in \mathbb{R}^m$. Since $C(\widehat{x})^{-1} \circ \Theta \circ C(\widehat{x}_0)$ preserves the splitting $\mathbb{R}^d \times \mathbb{R}^{m-d}$, the same holds for O. Also, diagonalizing R and estimating its eigenvalues, we get that if $\varepsilon > 0$ is small enough then $\|R - \mathrm{Id}\| \leq 4\sqrt{\varepsilon}$, see details in [**BO18**, p. 100]. Define $\delta = (\Psi^{-1}_{\widehat{x}} \circ \Psi_{\widehat{x}_0})(0) \in \mathbb{R}^m$, and $\Delta : B[10Q(\widehat{x}_0)] \to \mathbb{R}^m$ s.t. $\Psi^{-1}_{\widehat{x}} \circ \Psi_{\widehat{x}_0} = \delta + O + \Delta$. We start estimating $d\Delta$. For $z \in B[10Q(\widehat{x}_0)]$,

$$\begin{aligned}(d\Delta)_z &= C(\widehat{x})^{-1} \circ (d\Phi)_{C(\widehat{x}_0)z} \circ C(\widehat{x}_0) - O \\ &= C(\widehat{x})^{-1} \circ \underbrace{\left[(d\Phi)_{C(\widehat{x}_0)z} - \Theta\right]}_{=E} \circ C(\widehat{x}_0) + OR - O.\end{aligned}$$

To estimate E, observe that:

- By Lemma 6.3(2), $\|P - \Theta\| \leq \frac{1}{2}\eta^{15\beta/48}$.

∘ By assumptions (A2)–(A3),

$$\begin{aligned}
&\left\|(d\Phi)_{C(\widehat{x}_0)z}-P\right\|\\
&=\left\|\widetilde{(\mathrm{dexp}_x^{-1})_{\Psi_{\widehat{x}_0}(z)}(\mathrm{dexp}_{x_0})_{C(\widehat{x}_0)z}}-\widetilde{\mathrm{Id}}\right\|\\
&=\left\|\widetilde{(\mathrm{dexp}_x^{-1})_{\Psi_{\widehat{x}_0}(z)}(\mathrm{dexp}_{x_0})_{C(\widehat{x}_0)z}}-\widetilde{(\mathrm{dexp}_{x_0}^{-1})_{\Psi_{\widehat{x}_0}(z)}(\mathrm{dexp}_{x_0})_{C(\widehat{x}_0)z}}\right\|\\
&\le\left\|\widetilde{(\mathrm{dexp}_x^{-1})_{\Psi_{\widehat{x}_0}(z)}}-\widetilde{(\mathrm{dexp}_{x_0}^{-1})_{\Psi_{\widehat{x}_0}(z)}}\right\|\cdot\left\|(\mathrm{dexp}_{x_0})_{C(\widehat{x}_0)z}\right\|\\
&\le 2d(x_0,\mathscr{S})^{-a}d(x,x_0)
\end{aligned}$$

which, by part (1), is bounded by

$$\frac{2}{25}d(x_0,\mathscr{S})^{-a}\eta\le\frac{2}{25}\rho(\widehat{x}_0)^{-a}\eta\ll\frac{1}{2}\eta^{15\beta/48}.$$

Hence, $\|E\|<\eta^{15\beta/48}$ and so, by part (2), we get that

$$\begin{aligned}
\left\|C(\widehat{x})^{-1}\circ E\circ C(\widehat{x}_0)\right\|&\le\left\|C(\widehat{x})^{-1}\right\|\eta^{15\beta/48}\\
&\le e^{2\sqrt{\varepsilon}}\left\|C(\widehat{x}_0)^{-1}\right\|\eta^{15\beta/48}\\
&\le e^{2\sqrt{\varepsilon}}\varepsilon^{1/8}\eta^{14\beta/48}\ll\sqrt{\varepsilon}.
\end{aligned}$$

Since $\|OR-O\|=\|R-\mathrm{Id}\|\le 4\sqrt{\varepsilon}$, we conclude that $\|(d\Delta)_z\|\le 5\sqrt{\varepsilon}$. In particular, since $\Delta(0)=0$, we have $\|\Delta(z)\|\le\|d\Delta\|_{C^0}\|z\|\le 5\sqrt{\varepsilon}\|z\|$.

Finally, we estimate $\|\delta\|$. Let z s.t. $\Psi_{\widehat{x}_0}(z)=x$. We have $0=(\Psi_{\widehat{x}}^{-1}\circ\Psi_{\widehat{x}_0})(z)=\delta+Oz+\Delta(z)$ and so $\delta=-Oz-\Delta(z)$. By Lemma 4.4 we have $\|z\|<50^{-1}\eta$, therefore for $\varepsilon>0$ small enough

$$\|\delta\|\le\|Oz\|+\|\Delta(z)\|\le(1+5\sqrt{\varepsilon})\|z\|\le\tfrac{1+5\sqrt{\varepsilon}}{50}\eta<25^{-1}\eta.$$

6.8. Proof of Corollary 6.2

Parts (1)–(2) are consequences of parts (1) and (4) of Theorem 6.1. We prove part (3) as we proved Theorem 6.1(5), with the required modifications. Write $\pi(\underline{v})=\pi(\underline{w})=\widehat{x}$. Assume $n=0$, and let $x=\vartheta[\widehat{x}]$, $x_0=\vartheta[\widehat{x}_0]$, $y_0=\vartheta[\widehat{y}_0]$, $\eta=p_0^s\wedge p_0^u$, $\overline{\eta}=q_0^s\wedge q_0^u$. By Theorem 6.1(4), $\eta/\overline{\eta}=e^{\pm 2\sqrt[3]{\varepsilon}}$. Let

$$A=\Theta_{\widehat{y}_0,\widehat{x}}^{-1},\quad B=\Theta_{\widehat{x}_0,\widehat{x}},\quad\Theta=A\circ B$$

$$C=P_{y_0,x}^{-1},\quad D=P_{x_0,x},\quad P=C\circ D.$$

The proof that $\Psi_{\widehat{y}_0}^{-1}\circ\Psi_{\widehat{x}_0}$ is well-defined in $B[10Q(\widehat{x}_0)]$ is the same. Now, we have $\Psi_{\widehat{y}_0}^{-1}\circ\Psi_{\widehat{x}_0}=C(\widehat{y}_0)^{-1}\circ\Phi\circ C(\widehat{x}_0)$, where $\Phi=\exp_{y_0}^{-1}\circ\exp_{x_0}$, and $\|C(\widehat{y}_0)^{-1}\circ\Theta\circ C(\widehat{x}_0)v\|=e^{\pm 4\sqrt{\varepsilon}}\|v\|$ for all v. Hence $C(\widehat{y}_0)^{-1}\circ\Theta\circ C(\widehat{x}_0)=OR$ where O is an orthogonal matrix preserving the splitting $\mathbb{R}^d\times\mathbb{R}^{m-d}$ and R is positive symmetric with $\|R-\mathrm{Id}\|\le 8\sqrt{\varepsilon}$. Let $\delta=(\Psi_{\widehat{y}_0}^{-1}\circ\Psi_{\widehat{x}_0})(0)\in\mathbb{R}^m$ and Δ s.t. $\Psi_{\widehat{y}_0}^{-1}\circ\Psi_{\widehat{x}_0}=\delta+O+\Delta$. If $y\in B[10Q(\widehat{x}_0)]$ then

$$(d\Delta)_z=C(\widehat{y}_0)^{-1}\circ E\circ C(\widehat{x}_0)+OR-O$$

where $E=(d\Phi)_{C(\widehat{x}_0)z}-\Theta$. Note that:

- By Lemma 6.3(2), $\left\|A^{-1} - C^{-1}\right\| \le \frac{1}{2}\overline{\eta}^{15\beta/48}$ and $\|B - D\| \le \frac{1}{2}\eta^{15\beta/48}$. We also have $\|A\|, \|B\|, \|C\|, \|D\| < 2$. Since $A - C = A(C^{-1} - A^{-1})C$, we have

$$\|A - C\| \le 4\left\|A^{-1} - C^{-1}\right\| \le 2\overline{\eta}^{15\beta/48}$$

and so

$$\begin{aligned}\|P - \Theta\| &= \|AB - CD\| \\ &\le \|A\|\|B - D\| + \|A - C\|\|D\| \\ &\le \eta^{15\beta/48} + 4\overline{\eta}^{15\beta/48} \\ &\le (e^{2\sqrt[3]{\varepsilon}} + 4)\overline{\eta}^{15\beta/48} \\ &< 6\overline{\eta}^{15\beta/48}\end{aligned}$$

for $\varepsilon > 0$ small enough.

- Proceeding similarly as before and using part (1), we have

$$\begin{aligned}&\left\|(d\Phi)_{C(\widehat{x}_0)z} - P\right\| \le 2d(y_0, \mathscr{S})^{-a} d(x_0, y_0) \le 2\rho(\widehat{y}_0)^{-a}\tfrac{1}{10}\max\{\eta, \overline{\eta}\} \\ &\le \tfrac{1}{5}e^{2\sqrt[3]{\varepsilon}}\rho(\widehat{y}_0)^{-a}\overline{\eta} \ll \overline{\eta}^{15\beta/48}.\end{aligned}$$

Hence $\|E\| < 7\overline{\eta}^{15\beta/48}$ and so

$$\left\|C(\widehat{y}_0)^{-1} \circ E \circ C(\widehat{x}_0)\right\| \le 7\left\|C(\widehat{y}_0)^{-1}\right\|\overline{\eta}^{15\beta/48} \le 7\varepsilon^{1/8}\overline{\eta}^{14\beta/48} \ll \sqrt{\varepsilon},$$

which implies that $\|(d\Delta)_z\| < 9\sqrt{\varepsilon}$. To estimate $\|\delta\|$ we let z s.t. $\Psi_{\widehat{x}_0}(z) = x$. By Lemma 4.4 we have $\|z\| < 50^{-1}\eta$ and $\|(\Psi_{\widehat{y}_0}^{-1} \circ \Psi_{\widehat{x}_0})(z)\| < 50^{-1}\overline{\eta}$, and so

$$\begin{aligned}\|\delta\| &\le \|(\Psi_{\widehat{y}_0}^{-1} \circ \Psi_{\widehat{x}_0})(z)\| + \|Oz\| + \|\Delta(z)\| \\ &\le \|(\Psi_{\widehat{y}_0}^{-1} \circ \Psi_{\widehat{x}_0})(z)\| + (1 + 9\sqrt{\varepsilon})\|z\| \\ &< 50^{-1}\overline{\eta} + 50^{-1}(1 + 9\sqrt{\varepsilon})\eta \\ &\le 50^{-1}\left[1 + (1 + 9\sqrt{\varepsilon})e^{2\sqrt[3]{\varepsilon}}\right]\overline{\eta} \\ &< 10^{-1}\overline{\eta}\end{aligned}$$

for $\varepsilon > 0$ small enough.

CHAPTER 7

Symbolic dynamics

We are approaching the completion of the proof of Theorem 1.8. In this section, we will employ the same methods used in [**Sar13**, **BO18**] to our context (the results are stated in terms of $\widehat{M}$). This part is more abstract than the material of the previous sections, and whenever possible we will just invoke the results of [**Sar13**, **BO18**]. When needed, we will provide the complete details. In the sequel, we fix $\chi > 0$ and for $\varepsilon > 0$ small enough we let (Σ, σ, π) be the triple given by Proposition 5.2. Most of the results below will use Lemma 4.8, Proposition 4.9, Theorem 5.1, Proposition 5.2, and Corollary 6.2.

7.1. A countable Markov partition

We use Corollary 6.2 to construct a cover of $\mathrm{NUH}^{\#}$ that is locally finite and satisfies a (symbolic) Markov property.

THE MARKOV COVER $\mathscr{Z}$: Let $\mathscr{Z} := \{Z(v) : v \in \mathscr{A}\}$, where

$$Z(v) := \{\pi(\underline{v}) : \underline{v} \in \Sigma^{\#} \text{ and } v_0 = v\}.$$

In other words, $\mathscr{Z}$ is the family defined by the natural partition of $\Sigma^{\#}$ into cylinder at the zeroth position. Using stable/unstable sets, we define *invariant sets* inside each $Z \in \mathscr{Z}$. Let $Z = Z(v)$.

s/u-SETS IN $\mathscr{Z}$: Given $\widehat{x} \in Z$, let $\widehat{W}^s(\widehat{x}, Z) := \widehat{V}^s[\{v_n\}_{n\geq 0}] \cap Z$ be the *s-set* of $\widehat{x}$ in Z for some (any) $\underline{v} = \{v_n\}_{n\in\mathbb{Z}} \in \Sigma^{\#}$ s.t. $\pi(\underline{v}) = \widehat{x}$ and $v_0 = v$. Similarly, let $\widehat{W}^u(\widehat{x}, Z) := \widehat{V}^u[\{v_n\}_{n\leq 0}] \cap Z$ be the *u-set* of $\widehat{x}$ in Z.

By Proposition 4.9(5), the definitions above do not depend on the choice of $\underline{v}$, and any two s-sets (u-sets) either coincide or are disjoint. We also define $\widehat{V}^s(\widehat{x}, Z) := \widehat{V}^s[\{v_n\}_{n\geq 0}]$ and $\widehat{V}^u(\widehat{x}, Z) := \widehat{V}^u[\{v_n\}_{n\leq 0}]$. Here are the main properties of $\mathscr{Z}$.

PROPOSITION 7.1. *The following holds for all $\varepsilon > 0$ small enough.*

1. COVERING PROPERTY: $\bigcup_{Z\in\mathscr{Z}} Z = \mathrm{NUH}^{\#}$.
2. LOCAL FINITENESS: *For every $Z \in \mathscr{Z}$, $\#\{Z' \in \mathscr{Z} : Z \cap Z' \neq \emptyset\} < \infty$.*
3. PRODUCT STRUCTURE: *For every $Z \in \mathscr{Z}$ and every $\widehat{x}, \widehat{y} \in Z$, the intersection $\widehat{W}^s(\widehat{x}, Z) \cap \widehat{W}^u(\widehat{y}, Z)$ consists of a single point, and this point belongs to Z.*
4. SYMBOLIC MARKOV PROPERTY: *If $\widehat{x} = \pi(\underline{v})$ with $\underline{v} = \{v_n\}_{n\in\mathbb{Z}} \in \Sigma^{\#}$, then*

$$\widehat{f}(\widehat{W}^s(\widehat{x}, Z(v_0))) \subset \widehat{W}^s(\widehat{f}(\widehat{x}), Z(v_1)) \text{ and } \widehat{f}^{-1}(\widehat{W}^u(\widehat{f}(\widehat{x}), Z(v_1))) \subset \widehat{W}^u(\widehat{x}, Z(v_0)).$$

PROOF. By Proposition 5.2(4), we have $\pi[\Sigma^{\#}] \supset \mathrm{NUH}^{\#}$. On the other hand, Theorem 6.1 implies that $\pi[\Sigma^{\#}] \subset \mathrm{NUH}^{\#}$, hence part (1) follows. Part (2) is proved as in [**Sar13**, Theorem 10.2], using Corollary 6.2(2) and Theorem 5.1(1). Part (3) is proved as in [**Sar13**, Proposition 10.5], using Proposition 4.9(1). Part (4) is proved as in [**Sar13**, Proposition 10.9], using Lemma 4.8 and Proposition 4.9(3). □

For $\widehat{x}, \widehat{y} \in Z$, let $[\widehat{x}, \widehat{y}]_Z :=$ intersection point of $\widehat{W}^s(\widehat{x}, Z)$ and $\widehat{W}^u(\widehat{y}, Z)$, and call it the *Smale bracket* of $\widehat{x}, \widehat{y}$ in Z.

LEMMA 7.2. *The following holds for all $\varepsilon > 0$ small enough.*

(1) COMPATIBILITY: *If $\widehat{x}, \widehat{y} \in Z(v_0)$ and $\widehat{f}(\widehat{x}), \widehat{f}(\widehat{y}) \in Z(v_1)$ with $v_0 \overset{\varepsilon}{\to} v_1$ then $\widehat{f}([\widehat{x}, \widehat{y}]_{Z(v_0)}) = [\widehat{f}(\widehat{x}), \widehat{f}(\widehat{y})]_{Z(v_1)}$.*

(2) OVERLAPPING CHARTS PROPERTIES: *If $Z = Z(\Psi_{\widehat{x}_0}^{p_0^s, p_0^u})$, $Z' = Z(\Psi_{\widehat{y}_0}^{q_0^s, q_0^u}) \in \mathscr{Z}$ with $Z \cap Z' \neq \emptyset$ then:*
 (a) *$\vartheta[Z] \subset \Psi_{\widehat{y}_0}(B[q_0^s \wedge q_0^u])$.*
 (b) *If $\widehat{x} \in Z \cap Z'$ then $\widehat{W}^{s/u}(\widehat{x}, Z) \subset \widehat{V}^{s/u}(\widehat{x}, Z')$.*
 (c) *If $\widehat{x} \in Z, \widehat{y} \in Z'$ then $\widehat{V}^s(\widehat{x}, Z)$ and $\widehat{V}^u(\widehat{y}, Z')$ intersect at a unique point.*

PROOF. Part (1) is proved as in [**Sar13**, Lemma 10.7], using Proposition 4.9. The proof of part (2) is based on [**Sar13**, Lemmas 10.8 and 10.10] and [**BO18**, Lemmas 5.8 and 5.10]. Since the referred proofs are not simple applications of previous results, and since they use two norms (the euclidean and supremum norms), we will provide the complete details below.

Let $p = p_0^s \wedge p_0^u$ and $q = q_0^s \wedge q_0^u$. Since $Z \cap Z' \neq \emptyset$, Corollary 6.2 implies that $\frac{p}{q} = e^{\pm 2\sqrt[3]{\varepsilon}}$ and that $\Psi_{\widehat{y}_0}^{-1} \circ \Psi_{\widehat{x}_0} = \delta + O + \Delta$ on $B[10Q(\widehat{x}_0)]$, where $\delta \in \mathbb{R}^m$ satisfies $\|\delta\| < 10^{-1}q$, O is an orthogonal linear map preserving the splitting $\mathbb{R}^d \times \mathbb{R}^{m-d}$, and $\Delta : B[10Q(\widehat{x}_0)] \to \mathbb{R}^m$ satisfies $\Delta(0) = 0$ and $\|d\Delta\|_{C^0} < 9\sqrt{\varepsilon}$ on $B[10Q(\widehat{x}_0)]$. In particular, $\|\Delta(v)\| \leq 9\sqrt{\varepsilon}\|v\|$ for all $v \in B[10Q(\widehat{x}_0)]$.

(a) By Lemma 4.4 and the definition of Z, we have $\vartheta[Z] \subset \Psi_{\widehat{x}_0}(B[50^{-1}p])$. If $v \in B[50^{-1}p]$ then

$$\begin{aligned}\|(\Psi_{\widehat{y}_0}^{-1} \circ \Psi_{\widehat{x}_0})(v)\| &\leq \|\delta\| + \|Ov\| + \|\Delta(v)\| \leq \|\delta\| + (1 + 9\sqrt{\varepsilon})\|v\| \\ &\leq 10^{-1}q + (1 + 9\sqrt{\varepsilon})50^{-1}p \leq \left[10^{-1} + 50^{-1}(1 + 9\sqrt{\varepsilon})e^{2\sqrt[3]{\varepsilon}}\right] q < q\end{aligned}$$

for $\varepsilon > 0$ small enough, and so $\vartheta[Z] \subset \Psi_{\widehat{y}_0}[(\Psi_{\widehat{y}_0}^{-1} \circ \Psi_{\widehat{x}_0})(B[50^{-1}p])] \subset \Psi_{\widehat{y}_0}(B[q])$.

(b) Let $\underline{v} = \{v_n\}_{n\in\mathbb{Z}} = \{\Psi_{\widehat{x}_n}^{p_n^s, p_n^u}\}_{n\in\mathbb{Z}}$ and $\underline{w} = \{w_n\}_{n\in\mathbb{Z}} = \{\Psi_{\widehat{y}_n}^{q_n^s, q_n^u}\}_{n\in\mathbb{Z}}$ in $\Sigma^\#$ s.t. $\widehat{x} = \pi(\underline{v}) = \pi(\underline{w})$. We have to show that $\widehat{W}^{s/u}(\widehat{x}, Z) \subset \widehat{V}^{s/u}(\widehat{x}, Z')$. Let us start proving the statement for $\widehat{W}^s$. Fix $n \geq 0$. By Proposition 5.2(3), we have $\widehat{f}^n(\widehat{x}) = \pi[\sigma^n(\underline{v})] = \pi[\sigma^n(\underline{w})]$ and so $\widehat{f}^n(\widehat{x}) \in Z(v_n) \cap Z(w_n)$. By item (a) above, $\vartheta[Z(v_n)] \subset \Psi_{\widehat{y}_n}(B[q_n^s \wedge q_n^u]) \subset \Psi_{\widehat{y}_n}(B[20Q(\widehat{y}_n)])$. Now, Proposition 7.1(4) implies that $\widehat{f}^n(\widehat{W}^s(\widehat{x}, Z)) \subset \widehat{W}^s(\widehat{f}^n(\widehat{x}), Z(v_n)) \subset Z(v_n)$ and so

$$\vartheta_n\left[\widehat{W}^s(\widehat{x}, Z)\right] = \vartheta\left[\widehat{f}^n(\widehat{W}^s(\widehat{x}, Z))\right] \subset \vartheta[Z(v_n)] \subset \Psi_{\widehat{y}_n}(B[20Q(\widehat{y}_n)]).$$

By Lemma 4.8, it follows that $\widehat{W}^s(\widehat{x}, Z) \subset \widehat{V}^s[\{w_n\}_{n\geq 0}] = \widehat{V}^s(\widehat{x}, Z')$. The proof for $\widehat{W}^u$ is analogous, once we pay attention that the equality $\widehat{x} = \pi(\underline{v}) = \pi(\underline{w})$ implies that the inverse branches associated to $\underline{v}$ and $\underline{w}$ coincide.

(c) We apply methods very similar to those used in the appendix. Let $\widehat{V}^s(\widehat{x}, Z) = \widehat{V}^s[\underline{v}^+]$ and $\widehat{V}^u(\widehat{y}, Z') = \widehat{V}^u[\underline{w}^-]$. Any point in $\widehat{V}^s(\widehat{x}, Z) \cap \widehat{V}^u(\widehat{y}, Z')$ is defined by its zeroth position, so we need to show that $V^s[\underline{v}^+]$ and $V^u[\underline{w}^-]$ intersect at a single point. Write $V^s = V^s[\underline{v}^+]$, and let $G : B^d[p_0^s] \to \mathbb{R}^{m-d}$ be the representing function of V^s, i.e. $V^s = \Psi_{\widehat{x}_0}\{(v, G(v)) : v \in B^d[p_0^s]\}$. Recall that G satisfies (AM1)–(AM3) of Section 4.4. By item (a), we can express V^s in the chart $\Psi_{\widehat{y}_0}$ by

$$\begin{aligned} V^s &= \left[\Psi_{\widehat{y}_0} \circ (\Psi_{\widehat{y}_0}^{-1} \circ \Psi_{\widehat{x}_0})\right] \{(v, G(v)) : v \in B^d[p_0^s]\} \\ &= \Psi_{\widehat{y}_0} \left\{\delta + O(v, G(v)) + \Delta(v, G(v)) : v \in B^d[p_0^s]\right\}. \end{aligned}$$

We want to express $\delta + O(v, G(v)) + \Delta(v, G(v))$ as a graph. With respect to the splitting $\mathbb{R}^d \times \mathbb{R}^{m-d}$, write $\delta = (\delta_1, \delta_2)$ and $\Delta = (\Delta_1, \Delta_2)$. Since O preserves such splitting, if we write $w = Ov$ then

$$\begin{aligned} \delta + O(v, G(v)) + \Delta(v, G(v)) &= (Ov + \Delta_1(v, G(v)) + \delta_1, OG(v) + \Delta_2(v, G(v)) + \delta_2) \\ &= (w + \overline{\Delta}_1(w) + \delta_1, OGO^{-1}w + \overline{\Delta}_2(w) + \delta_2) \\ &= (\Gamma(w), OGO^{-1}w + \overline{\Delta}_2(w) + \delta_2), \end{aligned}$$

where $\overline{\Delta}_1(w) = \Delta_1(O^{-1}w, GO^{-1}w)$, $\overline{\Delta}_2(w) = \Delta_2(O^{-1}w, GO^{-1}w)$ and $\Gamma = \mathrm{Id} + \overline{\Delta}_1 + \delta_1$ are defined in $B^d[p_0^s]$. We start collecting some estimates for $\overline{\Delta}_1, \overline{\Delta}_2, \Gamma$:

- $\|\overline{\Delta}_i(0)\| = \|\Delta_i(0, G(0))\| \le \|d\Delta_i\|_{C^0}\|G(0)\| \le 9\sqrt{\varepsilon}10^{-3}p < 10^{-2}\sqrt{\varepsilon}p$.
- $\|d\overline{\Delta}_i\|_{C^0} \le \|d\Delta_i\|_{C^0}(1 + \|dG\|_{C^0}) < 9\sqrt{\varepsilon}(1+\varepsilon) < 10\sqrt{\varepsilon}$.
- $\|\Gamma(0)\| \le \|\overline{\Delta}_1(0)\| + \|\delta_1\| < 10^{-2}\sqrt{\varepsilon}p + 10^{-1}q < \frac{1}{6}(p \wedge q)$, since $\frac{p}{q} = e^{\pm 2\sqrt[3]{\varepsilon}}$.

In particular, if $v \in B^d[p_0^s]$ then

$$\|\overline{\Delta}_1(v)\| \le \|d\overline{\Delta}_1\|_{C^0}\|v\| + \|\overline{\Delta}_1(0)\| \le 10\sqrt{\varepsilon}p_0^s + 10^{-2}\sqrt{\varepsilon}p < 20\sqrt{\varepsilon}p_0^s$$

and so $\|\overline{\Delta}_1 + \delta_1\|_{C^0} \le 20\sqrt{\varepsilon}p_0^s + 10^{-1}q < 20\varepsilon^{3/2} + 10^{-1}\varepsilon < \varepsilon$. By Lemma A.1(2), Γ has an inverse $\Phi : \Gamma(B^d[p_0^s]) \to B^d[p_0^s]$. We can also estimate $\|d\Gamma\|_{C^0}$ and $\|d\Phi\|_{C^0}$ in their domains as definition, as follows:

- $d\Gamma = \mathrm{Id} + d\overline{\Delta}_1$ and so $\|d\Gamma\|_{C^0} = 1 \pm \|d\overline{\Delta}_1\|_{C^0} = 1 \pm 10\sqrt{\varepsilon} = e^{\pm 15\sqrt{\varepsilon}}$, since $e^{-15\sqrt{\varepsilon}} < 1 - 10\sqrt{\varepsilon} < 1 + 10\sqrt{\varepsilon} < e^{15\sqrt{\varepsilon}}$ for $\varepsilon > 0$ small enough.
- Letting $A = (d\overline{\Delta}_1)_v$, we have $(d\Phi)_{\Gamma(v)} = [(d\Gamma)_v]^{-1} = [\mathrm{Id} + A]^{-1} = \mathrm{Id} + \sum_{k\ge1}(-1)^kA^k$. Noting that $\left\|\sum_{k\ge1}(-1)^kA^k\right\| \le \sum_{k\ge1}\|A\|^k < \frac{10\sqrt{\varepsilon}}{1-10\sqrt{\varepsilon}} < 12\sqrt{\varepsilon}$, we conclude that $\|d\Phi\|_{C^0} = 1 \pm 12\sqrt{\varepsilon} = e^{\pm 15\sqrt{\varepsilon}}$, since $e^{-15\sqrt{\varepsilon}} < 1 - 12\sqrt{\varepsilon} < 1 + 12\sqrt{\varepsilon} < e^{15\sqrt{\varepsilon}}$ for $\varepsilon > 0$ small enough.

Introducing the variable $\widetilde{w} = \Gamma(w)$, we obtain that

$$V^s = \Psi_{\widehat{y}_0}\left\{(\widetilde{w}, \widetilde{G}(\widetilde{w})) : \widetilde{w} \in \Gamma(B^d[p_0^s])\right\},$$

where $\widetilde{G} : \Gamma(B^d[p_0^s]) \to \mathbb{R}^{m-d}$ is defined by $\widetilde{G} = (OGO^{-1} + \overline{\Delta}_2 + \delta_2) \circ \Phi$. Now we estimate $\|\widetilde{G}(0)\|$ and $\|d\widetilde{G}\|_{C^0}$. Since

$$\|\Phi(0)\| = \|\Phi(0) - \Phi(\Gamma(0))\| \le \|d\Phi\|_{C^0}\|\Gamma(0)\| \le e^{15\sqrt{\varepsilon}}\tfrac{1}{6}(p \wedge q)$$

and

$$\begin{aligned}\|OGO^{-1}\Phi(0)\| &= \|GO^{-1}\Phi(0)\| \\ &\le \|GO^{-1}\Phi(0) - G\underbrace{O^{-1}\Phi(\Gamma(0))}_{=0}\| + \|G(0)\| \\ &\le \|dG\|_{C^0}\|d\Phi\|_{C^0}\|\Gamma(0)\| + \|G(0)\| \\ &\le \varepsilon e^{15\sqrt{\varepsilon}}\frac{1}{6}(p\wedge q) + 10^{-3}p \\ &\le \left(\frac{1}{6}\varepsilon e^{15\sqrt{\varepsilon}} + 10^{-3}e^{2\sqrt[3]{\varepsilon}}\right)(p\wedge q) \\ &< 10^{-2}(p\wedge q),\end{aligned}$$

we obtain that

$$\begin{aligned}\|\widetilde{G}(0)\| &\le \|OGO^{-1}\Phi(0)\| + \|\overline{\Delta}_2(\Phi(0))\| + \|\delta_2\| \\ &\le \|OGO^{-1}\Phi(0)\| + \|d\overline{\Delta}_2\|_{C^0}\|\Phi(0)\| + \|\overline{\Delta}_2(0)\| + \|\delta_2\| \\ &\le 10^{-2}(p\wedge q) + 10\sqrt{\varepsilon}e^{15\sqrt{\varepsilon}}\tfrac{1}{6}(p\wedge q) + 10^{-2}\sqrt{\varepsilon}p + 10^{-1}q \\ &\le \left(10^{-2} + 2\sqrt{\varepsilon}e^{15\sqrt{\varepsilon}} + 10^{-2}\sqrt{\varepsilon}e^{2\sqrt[3]{\varepsilon}} + 10^{-1}e^{2\sqrt[3]{\varepsilon}}\right)(p\wedge q) \\ &< \frac{1}{6}(p\wedge q),\end{aligned}$$

where in the last inequality we used that $\varepsilon > 0$ is small enough. The estimate of $d\widetilde{G}$ is simpler:

$$\begin{aligned}\|d\widetilde{G}\|_{C^0} &\le \|O\circ dG\circ O^{-1} + d\overline{\Delta}_2\|_{C^0}\|d\Phi\|_{C^0} \\ &\le \left(\|dG\|_{C^0} + \|d\overline{\Delta}_2\|_{C^0}\right)\|d\Phi\|_{C^0} \\ &\le (\varepsilon + 10\sqrt{\varepsilon})e^{15\sqrt{\varepsilon}} < 15\sqrt{\varepsilon},\end{aligned}$$

where again in the last inequality we used that $\varepsilon > 0$ is small enough.

We finally prove item (c). We proceed similarly to the proof of Lemma A.10(1). Write $V^u = V^u[\widehat{w}^-]$, and let $H : B^{m-d}[q_0^u] \to \mathbb{R}^d$ be its representing function. The intersection $V^s \cap V^u$ corresponds to points $(\widetilde{w}, v)$ satisfying the system

$$\begin{cases} v = \widetilde{G}(\widetilde{w}) \\ \widetilde{w} = H(v) \end{cases}$$

for which necessarily $\widetilde{w}$ is a fixed point of $H\circ\widetilde{G}$. Note the following:

- If $\widetilde{w} \in B^d[\frac{2}{3}p]$ then $\|\widetilde{G}(\widetilde{w})\| \le \|d\widetilde{G}\|_{C^0}\|\widetilde{w}\| + \|\widetilde{G}(0)\| \le 15\sqrt{\varepsilon}\frac{2}{3}p + \frac{1}{6}(p\wedge q) < \frac{2}{3}p$.
- If $v \in B^{m-d}[\frac{2}{3}p]$ then $\|H(v)\| \le \|dH\|_{C^0}\|v\| + \|H(0)\| \le \varepsilon\frac{2}{3}p + 10^{-3}q < \frac{2}{3}p$.

Therefore

$$B^d[\tfrac{2}{3}p] \xrightarrow{\ \widetilde{G}\ } B^{m-d}[\tfrac{2}{3}p] \xrightarrow{\ H\ } B^d[\tfrac{2}{3}p],$$

so the restriction of $H\circ\widetilde{G}$ to $B^d[\frac{2}{3}p]$ is well-defined. The derivative of this composition is at most $\|dH\|_{C^0}\|d\widetilde{G}\|_{C^0} \le 15\varepsilon^{3/2} < 1$, so it has a fixed point $\widetilde{w} \in B^d[\frac{2}{3}p]$. For this fixed point, $v = \widetilde{G}(\widetilde{w}) \in B^{m-d}[\frac{2}{3}p] \subset B^{m-d}[q_0^u]$, and so we obtain a point in $V^s \cap V^u$. It remains to prove the uniqueness. Note that $\Gamma(B^d[p_0^s]) \subset B^d[2p_0^s]$. Leting $a = (2p_0^s) \vee q_0^u$, apply the Kirszbraun theorem to extend $\widetilde{G}$ and H to functions in the domains $B^d[a]$ and $B^{m-d}[a]$ with Lipschitz constants bounded by $15\sqrt{\varepsilon}$

and ε respectively. The same calculation above implies that

$$B^d[a] \xrightarrow{\widetilde{G}} B^{m-d}[a] \xrightarrow{H} B^d[a]$$

and that $H \circ \widetilde{G}$ is a contraction, hence $V^s \cap V^u$ consists of a single point. □

The next step is to refine $\mathscr{Z}$ to destroy its non-trivial intersections. The result will be a partition of $\mathrm{NUH}^\#$ by sets with the (geometrical) Markov property. This method, first developed by Sinaĭ and Bowen for finite covers [**Sin68a**, **Sin68b**, **Bow75**], also works for countable covers satisfying the local finiteness property [**Sar13**]. Write $\mathscr{Z} = \{Z_1, Z_2, \ldots\}$.

THE MARKOV PARTITION $\mathscr{R}$: For every $Z_i, Z_j \in \mathscr{Z}$, define a partition of Z_i by:

$$\begin{aligned}
T_{ij}^{su} &= \{\widehat{x} \in Z_i : \widehat{W}^s(\widehat{x}, Z_i) \cap Z_j \neq \emptyset, \widehat{W}^u(\widehat{x}, Z_i) \cap Z_j \neq \emptyset\},\\
T_{ij}^{s\emptyset} &= \{\widehat{x} \in Z_i : \widehat{W}^s(\widehat{x}, Z_i) \cap Z_j \neq \emptyset, \widehat{W}^u(\widehat{x}, Z_i) \cap Z_j = \emptyset\},\\
T_{ij}^{\emptyset u} &= \{\widehat{x} \in Z_i : \widehat{W}^s(\widehat{x}, Z_i) \cap Z_j = \emptyset, \widehat{W}^u(\widehat{x}, Z_i) \cap Z_j \neq \emptyset\},\\
T_{ij}^{\emptyset\emptyset} &= \{\widehat{x} \in Z_i : \widehat{W}^s(\widehat{x}, Z_i) \cap Z_j = \emptyset, \widehat{W}^u(\widehat{x}, Z_i) \cap Z_j = \emptyset\}.
\end{aligned}$$

Let $\mathscr{T} := \{T_{ij}^{\alpha\beta} : i, j \geq 1, \alpha \in \{s, \emptyset\}, \beta \in \{u, \emptyset\}\}$, and let $\mathscr{R}$ be the partition generated by $\mathscr{T}$. Clearly $T_{ii}^{su} = Z_i$, hence $\mathscr{R}$ is a partition of $\mathrm{NUH}^\#$. Using Lemma 7.2(2)(b), we prove that $T_{ij}^{su} = Z_i \cap Z_j$. Also, proceeding as in [**Sar13**, Proposition 11.2], $\mathscr{R}$ is the partition defined by the following equivalence relation in $\mathrm{NUH}^\#$:

$$\widehat{x} \sim \widehat{y} \iff \text{for all } Z, Z' \in \mathscr{Z} : \begin{cases} \widehat{x} \in Z & \Leftrightarrow \widehat{y} \in Z \\ \widehat{W}^s(\widehat{x}, Z) \cap Z' \neq \emptyset & \Leftrightarrow \widehat{W}^s(\widehat{y}, Z) \cap Z' \neq \emptyset \\ \widehat{W}^u(\widehat{x}, Z) \cap Z' \neq \emptyset & \Leftrightarrow \widehat{W}^u(\widehat{y}, Z) \cap Z' \neq \emptyset. \end{cases}$$

By definition, $\mathscr{R}$ is a refinement of $\mathscr{Z}$. By Corollary 6.2(2) and Theorem 5.1(1), $\mathscr{R}$ possesses two local finiteness properties:

- For every $Z \in \mathscr{Z}$, $\#\{R \in \mathscr{R} : R \subset Z\} < \infty$.
- For every $R \in \mathscr{R}$, $\#\{Z \in \mathscr{Z} : Z \supset R\} < \infty$.

Now we show that $\mathscr{R}$ is a Markov partition in the sense of Sinaĭ [**Sin68b**].

s/u-SETS IN $\mathscr{R}$: Given $\widehat{x} \in R \in \mathscr{R}$, we define the *$s$-set* and *$u$-set* of $\widehat{x}$ by:

$$\widehat{W}^s(\widehat{x}, R) := \bigcap_{\substack{T_{ij}^{\alpha\beta} \in \mathscr{T} \\ T_{ij}^{\alpha\beta} \supset R}} \widehat{W}^s(\widehat{x}, Z_i) \cap T_{ij}^{\alpha\beta} \text{ and } \widehat{W}^u(\widehat{x}, R) := \bigcap_{\substack{T_{ij}^{\alpha\beta} \in \mathscr{T} \\ T_{ij}^{\alpha\beta} \supset R}} \widehat{W}^u(\widehat{x}, Z_i) \cap T_{ij}^{\alpha\beta}.$$

We have that $\widehat{W}^{s/u}(\widehat{x}, R) \subset R$, $\widehat{W}^s(\widehat{x}, R) \cap \widehat{W}^u(\widehat{x}, R) = \{\widehat{x}\}$, and any two s-sets (u-sets) either coincide or are disjoint. The proofs are made as in [**Sar13**, Proposition 11.5(1)–(2)], using Proposition 4.9(5).

PROPOSITION 7.3. *The following are true.*

(1) PRODUCT STRUCTURE: *For every $R \in \mathscr{R}$ and every $\widehat{x}, \widehat{y} \in R$, the intersection $\widehat{W}^s(\widehat{x}, R) \cap \widehat{W}^u(\widehat{y}, R)$ consists of a single point, and this point belongs to R. Denote it by $[\widehat{x}, \widehat{y}]$.*

(2) HYPERBOLICITY: *If* $\widehat{y},\widehat{z}\in\widehat{W}^s(\widehat{x},R)$ *then* $d(\widehat{f}^n(\widehat{y}),\widehat{f}^n(\widehat{z}))\xrightarrow[n\to\infty]{}0$, *and if* $\widehat{y},\widehat{z}\in\widehat{W}^u(\widehat{x},R)$ *then* $d(\widehat{f}^n(\widehat{y}),\widehat{f}^n(\widehat{z}))\xrightarrow[n\to-\infty]{}0$. *The rates are exponential.*

(3) MARKOV PROPERTY: *Let* $R_0,R_1\in\mathscr{R}$. *If* $\widehat{x}\in R_0$ *and* $\widehat{f}(\widehat{x})\in R_1$ *then*

$$\widehat{f}(\widehat{W}^s(\widehat{x},R_0))\subset\widehat{W}^s(\widehat{f}(\widehat{x}),R_1)\ \text{ and }\ \widehat{f}^{-1}(\widehat{W}^u(\widehat{f}(x),R_1))\subset\widehat{W}^u(\widehat{x},R_0).$$

PROOF. Part (1) is proved as in [**Sar13**, Proposition 11.5(3)], using Lemma 7.2(2)(b). Along the proof we also obtain that $[\widehat{x},\widehat{y}]=[\widehat{x},\widehat{y}]_Z$ for every $Z\in\mathscr{Z}$ containing R. Part (2) is proved as in [**Sar13**, Proposition 11.5(4)], using Proposition 4.9(4). Finally, part (3) is proved as in [**Sar13**, Proposition 11.7], using Lemma 7.2(1), Proposition 7.1(4), Lemma 7.2(2)(b) and the equality $[\widehat{x},\widehat{y}]=[\widehat{x},\widehat{y}]_Z$ for all $\widehat{x},\widehat{y}\in R$ and all $Z\in\mathscr{Z}$ containing R. □

7.2. A finite-to-one Markov extension

Using the Markov partition $\mathscr{R}$, we now define a symbolic coding that satisfies Theorem 1.8. Let $\widehat{\mathscr{G}}=(\widehat{V},\widehat{E})$ be the oriented graph with vertices $\widehat{V}=\mathscr{R}$ and edges $\widehat{E}=\{R\to S: R,S\in\mathscr{R}\text{ s.t. }\widehat{f}(R)\cap S\neq\emptyset\}$, and let $(\widehat{\Sigma},\widehat{\sigma})$ be the TMS induced by $\widehat{\mathscr{G}}$. Since the ingoing and outgoing degrees of every vertex in Σ is finite, the same occurs to $\widehat{\Sigma}$, see [**Sar13**, Proposition 12.3].

For $\ell\in\mathbb{Z}$ and a path $R_m\to\cdots\to R_n$ on $\widehat{\mathscr{G}}$, define

$${}_\ell[R_m,\ldots,R_n]:=\widehat{f}^{-\ell}(R_m)\cap\cdots\cap\widehat{f}^{-\ell-(n-m)}(R_n),$$

that represents the set of points whose itinerary from ℓ to $\ell+(n-m)$ visits the rectangles $R_m,\ldots,R_n$. The crucial property that gives the new coding is that ${}_\ell[R_m,\ldots,R_n]\neq\emptyset$. This is a consequence of the Markov property of $\mathscr{R}$ (Proposition 7.3(3)), and the proof can be made by induction as in [**Sar13**, Lemma 12.1]. By Proposition 7.3(2), if $\{R_n\}_{n\in\mathbb{Z}}\in\widehat{\Sigma}$ then $\bigcap_{n\geq 0}\overline{{}_{-n}[R_{-n},\ldots,R_n]}$ is the intersection of a descending chain of nonempty closed sets with diameters converging to zero. We can therefore consider the following definition.

THE MAP $\widehat{\pi}:\widehat{\Sigma}\to\widehat{M}$: Given $\underline{R}=\{R_n\}_{n\in\mathbb{Z}}\in\widehat{\Sigma}$, $\widehat{\pi}(\underline{R})$ is defined by the identity

$$\{\widehat{\pi}(\underline{R})\}:=\bigcap_{n\geq 0}\overline{{}_{-n}[R_{-n},\ldots,R_n]}.$$

As we will prove, $(\widehat{\Sigma},\widehat{\sigma},\widehat{\pi})$ is a triple satisfying Theorem 1.8. To show this, we start relating π and $\widehat{\pi}$. Consider the cylinder sets in Σ: for $\ell\in\mathbb{Z}$ and a path $v_m\overset{\varepsilon}{\to}\cdots\overset{\varepsilon}{\to}v_n$ in Σ, let

$$Z_\ell[v_m,\ldots,v_n]:=\{\pi(\underline{w}):\underline{w}\in\Sigma^\#\text{ and }w_\ell=v_m,\ldots,w_{\ell+(n-m)}=v_n\}.$$

By Proposition 5.2(2), the diameter of $Z_{-n}[v_{-n},\ldots,v_n]$ goes to zero as $n\to\infty$.

LEMMA 7.4. *The following are true.*

(1) *For each* $\underline{R}=\{R_n\}_{n\in\mathbb{Z}}\in\widehat{\Sigma}$ *and* $Z(v)\supset R_0$, *there exists* $\underline{v}=\{v_n\}_{n\in\mathbb{Z}}\in\Sigma$ *with* $v_0=v$ *s.t.* ${}_m[R_m,\ldots,R_n]\subset Z_m[v_m,\ldots,v_n]$ *for every* $m\leq n$. *In particular,* $\widehat{\pi}(\underline{R})=\pi(\underline{v})$. *If* $\underline{R}\in\widehat{\Sigma}^\#$, *then* $\underline{v}\in\Sigma^\#$.

(2) *If* $R_m\to\cdots\to R_n$ *is a path on* $\widehat{\Sigma}$ *and* $v_m\overset{\varepsilon}{\to}\cdots\overset{\varepsilon}{\to}v_n$ *is a path on* Σ *s.t.* $R_k\subset Z(v_k)$ *for* $k=m,\ldots,n$, *then* ${}_m[R_m,\ldots,R_n]\subset Z_m[v_m,\ldots,v_n]$.

PROOF. The proof of part (1) follows [**BPVL20**, Claim on p. 20]. Fix $\underline{R} = \{R_n\}_{n\in\mathbb{Z}} \in \widehat{\Sigma}$ and v s.t. $Z(v) \supset R_0$. For each $k \geq 0$, fix some $\widehat{y}_k \in {}_{-k}[R_{-k}, \ldots, R_k]$. Since $\mathscr{R}$ and $\mathscr{Z}$ both cover the same set NUH$^{\#}$, there is $\underline{v}^{(k)} = \{v_\ell^{(k)}\}_{\ell\in\mathbb{Z}} \in \Sigma^{\#}$ with $v_0^{(k)} = v$ s.t. $\widehat{y}_k = \pi[\underline{v}^{(k)}]$. Since the degrees of the graph defining Σ are finite, for each $\ell \geq 0$ there are finitely many possibilities for the tuple $(v_{-\ell}^{(k)}, \ldots, v_\ell^{(k)})$. By a diagonal argument, there is $\underline{v} = \{v_\ell\}_{\ell\in\mathbb{Z}} \in \Sigma$ s.t. for each $\ell \geq 0$ we have $(v_{-\ell}, \ldots, v_\ell) = (v_{-\ell}^{(k)}, \ldots, v_\ell^{(k)})$ for infinitely many $k \geq 0$. We claim that $\underline{v}$ satisfies part (1). Clearly $v_0 = v$. For the other statements, firstly note that $\widehat{f}^n(\widehat{y}_k) = \pi[\sigma^n(\underline{v}^{(k)})]$, and so $R_n \subset Z_{v_n^{(k)}}$ for all $k \geq |n|$. In particular, $R_n \subset Z_{v_n}$. Now let $m \leq n$, and take $k \geq |m|, |n|$. Fix $\widehat{y} \in {}_m[R_m, \ldots, R_n]$. We need to show that $\widehat{y} \in Z_m[v_m, \ldots, v_n]$.

- Since $\widehat{f}^m(\widehat{y}) \in R_m \subset Z_{v_m}$, there is $\underline{w} \in \Sigma^{\#}$ with $w_m = v_m$ s.t. $\widehat{f}^m(\widehat{y}) = \pi[\sigma^m(\underline{w})]$.
- Since $\widehat{f}^n(\widehat{y}) \in R_n \subset Z_{v_n}$, there is $\underline{z} \in \Sigma^{\#}$ with $z_n = v_n$ s.t. $\widehat{f}^n(\widehat{y}) = \pi[\sigma^n(\underline{z})]$.

Define $\underline{a} = \{a_\ell\}_{\ell\in\mathbb{Z}}$ by:

$$a_\ell = \begin{cases} w_\ell & , \ell \leq m \\ v_\ell & , m \leq \ell \leq n \\ z_\ell & , \ell \geq n. \end{cases}$$

Since both $\underline{w}, \underline{z} \in \Sigma^{\#}$, also $\underline{a} \in \Sigma^{\#}$. Observe that:

- If $\ell \leq m$ then $\widehat{f}^\ell(\widehat{y}) = \pi[\sigma^\ell(\underline{w})] \in Z_{w_\ell} = Z_{a_\ell}$.
- If $m \leq \ell \leq n$ then $\widehat{f}^\ell(\widehat{y}) \in R_\ell \subset Z_{v_\ell} = Z_{a_\ell}$.
- If $\ell \geq n$ then $\widehat{f}^\ell(\widehat{y}) = \pi[\sigma^\ell(\underline{z})] \in Z_{z_\ell} = Z_{a_\ell}$.

Hence $\vartheta[\widehat{f}^\ell(\widehat{y})] \in \vartheta[Z_{a_\ell}]$ for all $\ell \in \mathbb{Z}$, which implies that $\underline{a}$ shadows $\widehat{y}$ and so $\pi(\underline{a}) = \widehat{y}$. In particular $\widehat{y} \in Z_m[v_m, \ldots, v_n]$. To prove the equality $\widehat{\pi}(\underline{R}) = \pi(\underline{v})$, firstly observe that

$$\{\widehat{\pi}(\underline{R})\} = \bigcap_{n\geq 0} \overline{{}_{-n}[R_{-n}, \ldots, R_n]} \subset \bigcap_{n\geq 0} \overline{Z_{-n}[v_{-n}, \ldots, v_n]}.$$

Since

$$\{\pi(\underline{v})\} = \bigcap_{n\geq 0} Z_{-n}[v_{-n}, \ldots, v_n] \subset \bigcap_{n\geq 0} \overline{Z_{-n}[v_{-n}, \ldots, v_n]},$$

the intersection $\bigcap_{n\geq 0} \overline{Z_{-n}[v_{-n}, \ldots, v_n]}$ is non-empty. But it consists of a descending chain of closed sets with diameter converging to zero, so it equals $\{\pi(\underline{v})\}$. It follows that $\widehat{\pi}(\underline{R}) = \pi(\underline{v})$. Finally, the local finiteness properties of $\mathscr{R}$ and $\mathscr{Z}$ imply that if $\underline{R} \in \widehat{\Sigma}^{\#}$ then $\underline{v} \in \Sigma^{\#}$. This concludes the proof of part (1).

The proof of part (2) follows exactly the same argument made above, when we took $\widehat{y} \in {}_m[R_m, \ldots, R_n]$ and proved that $\widehat{y} \in Z_m[v_m, \ldots, v_n]$. □

The non-injectivity of $\widehat{\pi}$ is analysed using the notion of affiliation [**Sar13**, §12.3].

AFFILIATION: Two rectangles $R, R' \in \mathscr{R}$ are called *affiliated* if there are $Z, Z' \in \mathscr{Z}$ s.t. $Z \supset R$, $Z' \supset R'$ and $Z \cap Z' \neq \emptyset$. If this occurs, we write $R \sim R'$.

LEMMA 7.5 (Bowen property). *If $\underline{R} = \{R_n\}_{n\in\mathbb{Z}}, \underline{S} = \{S_n\}_{n\in\mathbb{Z}} \in \widehat{\Sigma}^{\#}$, then $\widehat{\pi}(\underline{R}) = \widehat{\pi}(\underline{S})$ iff $R_n \sim S_n$ for all $n \in \mathbb{Z}$.*

PROOF.

($\Longrightarrow$) Assume that $\underline{R}, \underline{S} \in \widehat{\Sigma}^\#$ with $\widehat{x} = \widehat{\pi}(\underline{R}) = \widehat{\pi}(\underline{S})$. By Lemma 7.4, there are $\underline{v} = \{v_n\}_{n\in\mathbb{Z}}, \underline{w} = \{w_n\}_{n\in\mathbb{Z}} \in \Sigma^\#$ s.t. $\widehat{x} = \pi(\underline{v}) = \pi(\underline{w})$. By Proposition 5.2(3), for every $n \in \mathbb{Z}$ we have $\widehat{f}^n(\widehat{x}) = \pi[\sigma^n(\underline{v})] = \pi[\sigma^n(\underline{w})]$ and so $\widehat{f}^n(\widehat{x}) \in Z(v_n) \cap Z(w_n)$, proving that $R_n \sim S_n$.

($\Longleftarrow$) Assume that $\underline{R} = \{R_n\}_{n\in\mathbb{Z}}, \underline{S} = \{S_n\}_{n\in\mathbb{Z}} \in \widehat{\Sigma}^\#$ with $R_n \sim S_n$. Letting $\widehat{x} = \widehat{\pi}(\underline{R})$ and $\widehat{y} = \widehat{\pi}(\underline{S})$, we want to show that $\widehat{x} = \widehat{y}$. Start applying Lemma 7.4(1) to obtain $\underline{v} = \{v_n\}_{n\in\mathbb{Z}} = \{\Psi_{\widehat{x}_n}^{p_n^s, p_n^u}\} \in \Sigma^\#$ s.t. $\widehat{x} = \pi(\underline{v})$ and $\underline{w} = \{w_n\}_{n\in\mathbb{Z}} = \{\Psi_{\widehat{y}_n}^{q_n^s, q_n^u}\} \in \Sigma^\#$ s.t. $\widehat{y} = \pi(\underline{w})$. Had we that $Z(v_n) \cap Z(w_n) \neq \emptyset$ for all $n \in \mathbb{Z}$, we could just apply Lemma 7.2(2)(a) to obtain that $\widehat{x}$ is shadowed by $\underline{w}$, thus concluding that $\widehat{x} = \pi(\underline{w}) = \widehat{y}$. Unfortunately, we do not know if $Z(v_n)$ and $Z(w_n)$ intersect, but we do know that they cannot be far apart. Indeed, we claim that $\vartheta[Z(v_n)] \subset \Psi_{\widehat{y}_n}(B[20Q(\widehat{y}_n)])$ for all $n \in \mathbb{Z}$. Once this is proved, we get that $\widehat{x} = \widehat{y}$. We prove the inclusion for $n = 0$. Since $R_0 \sim S_0$, there are $a = \Psi_{\widehat{c}}^{r^s, r^u}$ and $b = \Psi_{\widehat{d}}^{t^s, t^u}$ s.t. $Z(a) \supset R_0$, $Z(b) \supset S_0$ and $Z(a) \cap Z(b) \neq \emptyset$. Since $Z(v_0), Z(a) \supset R_0$, we have that $Z(v_0) \cap Z(a) \neq \emptyset$. Similarly, $Z(w_0) \cap Z(b) \neq \emptyset$. Letting $r = r^s \wedge r^u$, $t = t^s \wedge t^u$ and $q = q_0^s \wedge q_0^u$, the same arguments used the proof of Lemma 7.2(2)(a) imply that

$$\vartheta[Z(v_0)] \subset \Psi_{\widehat{c}}(B[r]) \subset \Psi_{\widehat{d}}(B[3t]) \subset \Psi_{\widehat{y}_0}(B[9q]),$$

which is contained in $\Psi_{\widehat{y}_0}(B[20Q(\widehat{y}_0)])$. □

The next theorem completes the proof of Theorem 1.8.

THEOREM 7.6. *The following holds for all $\varepsilon > 0$ small enough.*

(1) $\widehat{\pi} : \widehat{\Sigma} \to \widehat{M}$ *is Hölder continuous.*
(2) $\widehat{\pi} \circ \widehat{\sigma} = \widehat{f} \circ \widehat{\pi}$.
(3) $\widehat{\pi} \restriction_{\widehat{\Sigma}^\#} : \widehat{\Sigma}^\# \to \mathrm{NUH}^\#$ *is a finite-to-one surjective map.*

PROOF.

(1) We have to show that the diameter of ${}_{-n}[R_{-n}, \ldots, R_n]$ decays exponentially fast in n. By Proposition 5.2(2), there are $C > 0$ and $\theta \in (0, 1)$ s.t. the diameter of every $Z_{-n}[v_{-n}, \ldots, v_n]$ is at most $C\theta^n$. For a fixed ${}_{-n}[R_{-n}, \ldots, R_n]$, apply Lemma 7.4 to obtain $v_{-n} \overset{\varepsilon}{\to} \cdots \overset{\varepsilon}{\to} v_n$ s.t. ${}_{-n}[R_{-n}, \ldots, R_n] \subset Z_{-n}[v_{-n}, \ldots, v_n]$. This concludes the proof.

(2) Fix $\underline{R} \in \widehat{\Sigma}$, and let $\underline{v} \in \Sigma$ satisfying Lemma 7.4(1). Then $\widehat{\pi}(\underline{R}) = \pi(\underline{v})$ and $\widehat{\pi}[\widehat{\sigma}(\underline{R})] = \pi[\sigma(\underline{v})]$. By Proposition 5.2(3), it follows that

$$\widehat{\pi}[\widehat{\sigma}(\underline{R})] = \pi[\sigma(\underline{v})] = \widehat{f}[\pi(\underline{v})] = \widehat{f}[\widehat{\pi}(\underline{R})].$$

(3) The proof is an adaptation of [**LS19**, Theorem 5.6(5)] to our context. We provide the details. For $R \in \mathscr{R}$, define

$$N(R) := \#\{(R', v') \in \mathscr{R} \times \mathscr{A} : R' \text{ is affiliated to } R \text{ and } Z(v') \supset R'\}.$$

By Proposition 7.1(2), $N(R)$ is finite. If $R, S \in \mathscr{R}$ then $N(R, S) := N(R)N(S)$ is also finite.

CLAIM: If $\widehat{x} = \widehat{\pi}(\underline{R})$ where $R_n = R$ for infinitely many $n < 0$ and $R_n = S$ for infinitely many $n > 0$, then $\#[\widehat{\pi}^{-1}(\widehat{x}) \cap \widehat{\Sigma}^\#] \leq N(R, S)$.

PROOF OF THE CLAIM. The proof is by contradiction. Write $N := N(R,S)$, and assume that $\underline{R}^{(0)},\dots,\underline{R}^{(N)} \in \widehat{\Sigma}^{\#}$ are distinct and satisfy $\widehat{\pi}[\underline{R}^{(j)}] = \widehat{x}$ for $j=0,\dots,N$. Write $\underline{R}^{(0)} = \underline{R}$ and $\underline{R}^{(j)} = \{R_k^{(j)}\}_{k\in\mathbb{Z}}$. By Lemma 7.4(1), for each j there is $\underline{v}^{(j)} \in \Sigma^{\#}$ with $\widehat{x} = \pi(\underline{v}^{(j)})$. By Lemma 7.5, for every $i \in \mathbb{Z}$ the sets $R_i^{(0)},\dots,R_i^{(N)}$ are affiliated. Fix n large enough s.t. the sequences $(R_{-n}^{(j)},\dots,R_n^{(j)})$, $j=0,\dots,N$, are all distinct, and fix $k,\ell > n$ s.t. $R_{-k}^{(0)} = R$ and $R_\ell^{(0)} = S$. There are at most $N = N(R)N(S)$ possibilities for the quadruple $(R_{-k}^{(j)}, v_{-k}^{(j)}; R_\ell^{(j)}, v_\ell^{(j)})$, $j=0,\dots,N$, and so by the pigeonhole principle there are $0 \le j_1, j_2 \le N$ with $j_1 \neq j_2$ s.t.

$$(R_{-k}^{(j_1)}, v_{-k}^{(j_1)}, R_\ell^{(j_1)}, v_\ell^{(j_1)}) = (R_{-k}^{(j_2)}, v_{-k}^{(j_2)}, R_\ell^{(j_2)}, v_\ell^{(j_2)}).$$

Write $\underline{A} := \underline{R}^{(j_1)}$, $\underline{B} := \underline{R}^{(j_2)}$, $\underline{a} := \underline{v}^{(j_1)}$, $\underline{b} := \underline{v}^{(j_2)}$, $A := A_\ell = B_\ell$, $B := A_{-k} = B_{-k}$, $a := a_\ell = b_\ell$ and $b := a_{-k} = b_{-k}$. Fix $\widehat{x}_A \in {}_{-k}[A_{-k},\dots,A_\ell]$ and $\widehat{x}_B \in {}_{-k}[B_{-k},\dots,B_\ell]$, and define $\widehat{z}_A, \widehat{z}_B$ by the equalities

$$\widehat{f}^{\ell}(\widehat{z}_A) := [\widehat{f}^{\ell}(\widehat{x}_A), \widehat{f}^{\ell}(\widehat{x}_B)] \text{ and } \widehat{f}^{-k}(\widehat{z}_B) := [\widehat{f}^{-k}(\widehat{x}_A), \widehat{f}^{-k}(\widehat{x}_B)].$$

These points are uniquely defined by Proposition 7.3(1), since $\widehat{f}^{\ell}(\widehat{x}_A), \widehat{f}^{\ell}(\widehat{x}_B) \in A$ and $\widehat{f}^{-k}(\widehat{x}_A), \widehat{f}^{-k}(\widehat{x}_B) \in B$. We will obtain a contradiction once we show that $\widehat{z}_A \neq \widehat{z}_B$ and $\widehat{z}_A = \widehat{z}_B$. Firstly, note that by Proposition 7.3(3) we have $\widehat{z}_A \in {}_{-k}[B_{-k},\dots,B_\ell]$ and $\widehat{z}_B \in {}_{-k}[A_{-k},\dots,A_\ell]$. Since $(A_{-k},\dots,A_\ell) \neq (B_{-k},\dots,B_\ell)$, we get that $\widehat{z}_A \neq \widehat{z}_B$.

Now we show that $\widehat{z}_A = \widehat{z}_B$. Since $\widehat{f}^{\ell}(\widehat{z}_A) \in A_\ell \subset Z(a)$ and $\widehat{f}^{-k}(\widehat{z}_B) \in B_{-k} \subset Z(b)$, there are $\underline{\alpha} = \{\alpha_n\}_{n\in\mathbb{Z}}, \underline{\beta} = \{\beta_n\}_{n\in\mathbb{Z}} \in \Sigma^{\#}$ with $\alpha_\ell = a$ and $\beta_{-k} = b$ s.t. $\widehat{z}_A = \pi(\underline{\alpha})$ and $\widehat{z}_B = \pi(\underline{\beta})$. Define $\underline{c} = \{c_n\}_{n\in\mathbb{Z}} \in \Sigma$ by:

$$c_n = \begin{cases} \beta_n & , n \le -k \\ a_n & , -k \le n \le \ell \\ \alpha_n & , n \ge \ell. \end{cases}$$

Since both $\underline{\alpha}, \underline{\beta} \in \Sigma^{\#}$, also $\underline{c} \in \Sigma^{\#}$. We claim that $\widehat{z}_A = \pi(\underline{c}) = \widehat{z}_B$. Write $c_n = \Psi_{\widehat{x}_n}^{p_n^s, p_n^u}$. By Proposition 7.3(3), $\widehat{f}^{-k}(\widehat{z}_A), \widehat{f}^{-k}(\widehat{z}_B)$ both belong to $\widehat{W}^u(\widehat{f}^{-k}(\widehat{x}_B), B)$. Since this latter s-set is contained in $\widehat{V}^u[\{c_n\}_{n\le -k}]$, Lemma 4.8 implies that

$$\vartheta_n[\widehat{z}_A], \vartheta_n[\widehat{z}_B] \in \Psi_{\widehat{x}_n}(B[20Q(\widehat{x}_n)]) \text{ for all } n \le -k.$$

Similarly, we obtain that

$$\vartheta_n[\widehat{z}_A], \vartheta_n[\widehat{z}_B] \in \Psi_{\widehat{x}_n}(B[20Q(\widehat{x}_n)]) \text{ for all } n \ge \ell.$$

Now fix $-k \le n \le \ell$. We have $\widehat{f}^n(\widehat{z}_A), \widehat{f}^n(\widehat{z}_B) \in A_n \cup B_n \subset Z(a_n) \cup Z(b_n)$. Since $\pi(\underline{a}) = \widehat{x} = \pi(\underline{b})$, we have $Z(a_n) \cap Z(b_n) \neq \emptyset$ and so, recalling that $a_n = c_n$, Lemma 7.2(2)(a) implies that $\vartheta[Z(a_n) \cup Z(b_n)] \subset \Psi_{\widehat{x}_n}(B[20Q(\widehat{x}_n)])$. Therefore

$$\vartheta_n[\widehat{z}_A], \vartheta_n[\widehat{z}_B] \in \Psi_{\widehat{x}_n}(B[20Q(\widehat{x}_n)]) \text{ for all } -k \le n \le \ell.$$

But then $\underline{c}$ shadows both $\widehat{z}_A$ and $\widehat{z}_B$ which, by Lemma 4.10, implies that $\widehat{z}_A = \widehat{z}_B$. We therefore obtain a contradiction. □

The claim concludes the proof of the theorem. □

We finish this section stating a result that will be used in Part 2. Since $\widehat{\pi}[\widehat{\Sigma}^\#] = \mathrm{NUH}^\#$ and every $\widehat{x} \in \mathrm{NUH}^\#$ has invariant subspaces $E^s_{\widehat{x}}, E^u_{\widehat{x}}$, each $\underline{R} \in \widehat{\Sigma}^\#$ is associated to subspaces $E^s_{\widehat{\pi}(\underline{R})}, E^u_{\widehat{\pi}(\underline{R})}$.

PROPOSITION 7.7. *The maps* $\underline{R} \in \widehat{\Sigma}^\# \mapsto E^{s/u}_{\widehat{\pi}(\underline{R})}$ *are Hölder continuous.*

The proof is made as in [**Sar13**, Lemma 12.6], using Lemma 7.4(2) and Proposition 4.7(5).

Part 2

Applications

In this part, we provide some applications of Theorem 1.8:

- Estimating the number of closed orbits.
- Establishing ergodic properties of equilibrium measures.

Let f satisfying conditions (A1)–(A7), and recall that $\vartheta : \widehat{M} \to M$ is the projection into the zeroth coordinate, see Section 2.2. Recall also that there are bijective projection/lift procedures between f-invariant probability measures μ in M and $\widehat{f}$-invariant probability measures $\widehat{\mu}$ in $\widehat{M}$. Using this bijection, call $\widehat{\mu}$ a χ-hyperbolic measure if μ is χ-hyperbolic, and $\widehat{\mu}$ an $\widehat{f}$-adapted measure if μ is f-adapted. In the sequel, let $\mathrm{NUH}^{\#}_{\chi}$ represent the set $\mathrm{NUH}^{\#}$ for the parameter $\chi > 0$. Also, let (Σ, σ, π) be the triple given by Theorem 1.8 for f and the parameter χ. The next result is of great relevance to our applications.

PROPOSITION 7.8. *Fix f satisfying conditions (A1)–(A7) and $\chi > 0$. Each f-adapted χ-hyperbolic measure μ in M can be lifted to a σ-invariant probability measure ν in Σ s.t.:*

(1) ERGODICITY: *μ is ergodic iff ν is ergodic.*
(2) ENTROPY: *$h_\mu(f) = h_\nu(\sigma)$.*
(3) INTEGRAL: *For all $\psi : M \to \mathbb{R}$ measurable it holds*

$$\int_M \psi d\mu = \int_\Sigma (\psi \circ \vartheta \circ \pi) d\nu.$$

Reversely, each σ-invariant probability measure ν in Σ projects to a χ-hyperbolic probability measure μ in M s.t. (1)–(3) hold.

Note that we do not claim that the projected measure μ is f-adapted.

PROOF. The bijection of Subsection 2.2 between f-invariant probability measures μ in M and $\widehat{f}$-invariant probability measures $\widehat{\mu}$ in $\widehat{M}$ preserve (1)–(3), and restricts to a bijection between f-adapted χ-hyperbolic measures in M and $\widehat{f}$-adapted χ-hyperbolic measures in $\widehat{M}$. By Lemma 3.6, if μ is f-adapted and χ-hyperbolic, then $\widehat{\mu}$ is carried by $\mathrm{NUH}^{\#}_{\chi}$. Proceeding as in [**Sar13**, §13], we can lift $\widehat{\mu}$ to a σ-invariant probability measure ν on Σ that satisfies (1)–(3).

Now we project measures. By the Poincaré recurrence theorem, every σ-invariant probability measure ν in Σ is carried by $\Sigma^{\#}$, hence it descends to an $\widehat{f}$-invariant probability measure $\widehat{\mu}$ carried by $\mathrm{NUH}^{\#}_{\chi}$ (attention: $\widehat{\mu}$ is not necessarily $\widehat{f}$-adapted). This projection clearly satisfies (1) and (3) and, since $\pi \restriction_{\Sigma^{\#}} : \Sigma^{\#} \to \mathrm{NUH}^{\#}_{\chi}$ is finite-to-one, it also satisfies (2). By the definition of $\mathrm{NUH}^{\#}_{\chi}$, $\widehat{\mu}$ descends to a χ-hyperbolic measure μ on M. □

CHAPTER 8

Equilibrium measures

Call a measure *hyperbolic* if it is χ-hyperbolic for some $\chi > 0$. We will restrict our attention to equilibrium measures of Hölder continuous potentials or of multiples of the geometric potential. Let us first recall some definitions. Let (Y, S), where Y is a complete metric separable space and $S : Y \to Y$ is a map, and let $\psi : Y \to \mathbb{R}$ be a (not necessarily continuous) potential.

TOPOLOGICAL PRESSURE: The *topological pressure* of ψ is $P_{\rm top}(\psi) := \sup\{h_\mu(S) + \int \psi d\mu\}$, where the supremum ranges over all S-invariant probability measures for which $\int \psi d\mu$ makes sense and $h_\mu(S) + \int \psi d\mu \neq \infty - \infty$.

EQUILIBRIUM MEASURE: An *equilibrium measure* for ψ is an S-invariant probability measure μ for which $\int \psi d\mu$ makes sense, $h_\mu(S) + \int \psi d\mu \neq \infty - \infty$, and $P_{\rm top}(\psi) = h_\mu(S) + \int \psi d\mu$.

When $\psi \equiv 0$, equilibrium measures are called *measures of maximal entropy* and $P_{\rm top}(0)$ is called the *topological entropy* of S, which we denote by $h_{\rm top}(S)$. If $\pi : (X, T) \to (Y, S)$ is finite-to-one, then equilibrium measures for ψ lift to equilibrium measures for $\varphi = \psi \circ \pi$. Note that φ is Hölder continuous whenever ψ, π are.

8.1. Hölder continuous potentials

In this section we prove the following result.

THEOREM 8.1. *Assume that f satisfies assumptions (A1)–(A7). Every equilibrium measure of a bounded Hölder continuous potential with finite topological pressure has at most countably many f-adapted hyperbolic ergodic components. Furthermore, the lift to $\widehat{M}$ of each such ergodic component is Bernoulli up to a period.*

Of course, if f is invertible then each measure itself is Bernoulli up to a period. Note that if $P_{\rm top}(0) < \infty$ then every bounded Hölder continuous potential has finite topological pressure.

PROOF. Let $\psi : M \to \mathbb{R}$ be a bounded Hölder continuous potential with $P_{\rm top}(\psi) < \infty$. We prove that, for each $\chi > 0$, ψ possesses at most countably many f-adapted χ-hyperbolic ergodic equilibrium measures. The first part of the theorem will follow by taking the union of these measures for $\chi_n = \frac{1}{n}$, $n > 0$. Fix $\chi > 0$, and let $\pi : \Sigma \to \widehat{M}$ be the coding given by Theorem 1.8 for the parameter χ. Define $\varphi : \Sigma \to \mathbb{R}$ by $\varphi = \psi \circ \vartheta \circ \pi$, a bounded Hölder continuous potential. We claim that $P_{\rm top}(\varphi) \leq P_{\rm top}(\psi)$. Indeed, for each σ-invariant probability measure ν there is an f-invariant probability measure μ satisfying Proposition 7.8, and so $h_\nu(\sigma) + \int \varphi d\nu = h_\mu(f) + \int \psi d\mu$. Taking the supremum over all ν, we conclude that $P_{\rm top}(\varphi) \leq P_{\rm top}(\psi)$. Now assume that μ is an f-adapted χ-hyperbolic equilibrium measure for ψ. Again by Proposition 7.8, there is a σ-invariant probability measure ν s.t. $h_\nu(\sigma)+\int \varphi d\nu = h_\mu(f)+\int \psi d\mu = P_{\rm top}(\psi)$, and so $P_{\rm top}(\varphi) \geq P_{\rm top}(\psi)$. Therefore, $P_{\rm top}(\varphi) = P_{\rm top}(\psi)$ and ν is an equilibrium measure for φ. By [**BS03**, Thm. 1.1], each topologically transitive countable topological Markov shift carries at most one equilibrium measure for φ. Hence, there are at most countably many such ν, which implies that there are at most countably many f-adapted χ-hyperbolic equilibrium measure for ψ. This concludes the proof of the first part of theorem.

Now we prove the second part. Buzzi & Sarig have shown how to construct, when they exist, equilibrium measures for bounded Hölder continuous potentials in symbolic spaces, see [**BS03**, Thm. 1.2]. Using this, Sarig proved that such measures have local product structure and are Bernoulli up to a period, see [**Sar11**]. Since these results are proved at a symbolic level, each ν in the previous paragraph is Bernoulli up to a period. Finally, since this latter property is preserved by finite-to-one factor maps, we conclude that $\widehat{\mu}$ is Bernoulli up to a period. $\square$

Theorem 8.1 also holds for unbounded Hölder continuous potentials, provided $\sup \psi < \infty$ and $P_{\rm top}(\psi) < \infty$. Indeed, these are the conditions required for the potential in [**BS03**, Thm. 1.1].

COROLLARY 8.2. *If f satisfies assumptions (A1)–(A7) and $h_{\rm top}(f) < \infty$, then f possesses at most countably many f-adapted hyperbolic ergodic measures of maximal entropy, and the lift to $\widehat{M}$ of each of them is Bernoulli up to a period.*

Again, if f is invertible then each measure itself is Bernoulli up to a period.

8.2. Multiples of the geometric potential

The proof of Theorem 8.1 relies on the properties on φ that allow to use [**BS03**]. We will prove, using Proposition 7.7, that these conditions are also satisfied for multiples of the geometric potential. We start with some definitions.

HYPERBOLIC LOCUS OF f: The *hyperbolic locus* of f is the set $\widehat{M}_{\rm hyp}$ defined as

$$\widehat{M}_{\rm hyp} := \bigcup_{\chi>0} {\rm NUH}_\chi.$$

Clearly, $\widehat{M}_{\rm hyp}$ is $\widehat{f}$-invariant and carries all hyperbolic measures. Observe that, by Proposition 4.11, each $\widehat{x} \in \widehat{M}_{\rm hyp}$ is associated to two subspaces $E^s_{\widehat{x}}, E^u_{\widehat{x}}$ of $\widehat{TM}_{\widehat{x}}$.

GEOMETRIC POTENTIAL OF f: The *geometric potential* of f is the function $J : \widehat{M}_{\rm hyp} \to \mathbb{R}$ defined by

$$J(\widehat{x}) := -\log\det\left[\widehat{df}\restriction_{E^u_{\widehat{x}}}\right].$$

Note that J is defined in the natural extension only. Indeed, it is in this set that unstable directions are well-defined. Let $t \in \mathbb{R}$.

TOPOLOGICAL PRESSURE OF tJ: The *topological pressure* of tJ is defined as $P_{\rm top}(tJ) = \sup\{h_{\widehat{\mu}}(\widehat{f}) + \int tJd\widehat{\mu}\}$ where the supremum ranges over all $\widehat{f}$-invariant probability measures $\widehat{\mu}$ supported on $\widehat{M}_{\rm hyp}$ for which $\int Jd\widehat{\mu}$ makes sense and $h_{\widehat{\mu}}(\widehat{f})+\int tJd\widehat{\mu} \neq \infty - \infty$.

EQUILIBRIUM MEASURE FOR tJ: An *equilibrium measure* for tJ is an $\widehat{f}$-invariant probability measure $\widehat{\mu}$ supported on $\widehat{M}_{\rm hyp}$ for which $\int Jd\widehat{\mu}$ makes sense, $h_{\widehat{\mu}}(\widehat{f})+\int tJd\widehat{\mu}\neq\infty-\infty$, and $P_{\rm top}(tJ)=h_{\widehat{\mu}}(\widehat{f})+\int tJd\widehat{\mu}$.

In the generality of (A1)–(A7), the functions tJ can be unbounded (possibly approaching $\pm\infty$), as well as $P_{\rm top}(tJ)$ might not be finite. Except for these possibilities, an analogue of Theorem 8.1 holds.

THEOREM 8.3. *Assume that f satisfies assumptions (A1)–(A7), and fix $t\in\mathbb{R}$. If $\sup(tJ)<\infty$ and $P_{\rm top}(tJ)<\infty$, then every equilibrium measure of tJ has at most countably many $\widehat{f}$-adapted hyperbolic ergodic components. Furthermore, each such ergodic component is Bernoulli up to a period.*

PROOF. Each ergodic hyperbolic equilibrium measure of tJ is χ-hyperbolic for some $\chi>0$. For each $\chi>0$, consider (Σ,σ,π) given by Theorem 1.8 and let $\varphi:\Sigma\to\mathbb{R}$ be defined by $\varphi=(tJ)\circ\pi$. Observe that φ is globally defined with $\sup(\varphi)<\infty$, and by Proposition 7.7 it is Hölder continuous. Proceeding as in the proof of Theorem 8.1, we get that $P_{\rm top}(\varphi)\leq P_{\rm top}(tJ)<\infty$. Hence we can apply [**BS03**] to φ and repeat the rest of the proof of Theorem 8.1. □

It is important noticing that, in our setting, neither the Ruelle inequality nor the Pesin entropy formula are known to hold, hence we do not know if an invariant Liouville measure is an equilibrium measure of J.

CHAPTER 9

Exponential growth rate of periodic orbits

Let f satisfying assumptions (A1)–(A7). For $\chi > 0$ and $n \geq 1$, let $\mathrm{Per}_{n,\chi}(f)$ denote the number of periodic points of period n for f with all Lyapunov exponents greater than χ in absolute value. The next result provides sufficient conditions for the exponential growth rate of $\mathrm{Per}_{n,\chi}(f)$.

THEOREM 9.1. *Assume that f satisfies assumptions (A1)–(A7). If f possesses an f-adapted χ-hyperbolic mixing measure with entropy $h > 0$, then*

$$\liminf_{n\to\infty} e^{-hn}\,\mathrm{Per}_{n,\chi}(f) \geq 1.$$

PROOF. This proof was suggested to us by Jérôme Buzzi, making use of his recent work [**Buz20**]. We will use some standard terminology of symbolic dynamics on Markov shifts, see e.g. [**Kit98**]. Let μ be f-adapted χ-hyperbolic mixing with $h_\mu(f) = h > 0$. Being mixing, μ is in particular ergodic. Consider the triple (Σ,σ,π) given by Theorem 1.8. Then $\pi\restriction_{\Sigma^\#}: \Sigma^\# \to \mathrm{NUH}^\#_\chi$ satisfies the Main Theorem of [**Buz20**]: it is a Borel semiconjugacy with the Bowen property (see Lemma 7.5). Hence there is a TMS $(\widetilde{\Sigma},\widetilde{\sigma})$ and a 1–Lipschitz map $\theta:\widetilde{\Sigma}\to\Sigma^\#$ s.t. $\pi\circ\theta:\widetilde{\Sigma}\to\mathrm{NUH}^\#_\chi$ is an *injective* semiconjugacy and $(\pi\circ\theta)[\widetilde{\Sigma}^\#]$ carries all $\widehat{f}$-invariant measures of $\pi[\Sigma^\#]=\mathrm{NUH}^\#_\chi$. This allows to lift $\widehat{\mu}$ to a *mixing* $\widetilde{\sigma}$-invariant probability measure λ on $\widetilde{\Sigma}$ with entropy h. If $\widetilde{\Sigma}$ is defined by the oriented graph $(\widetilde{V},\widetilde{E})$, then λ is carried by the topological Markov subshift $(\widetilde{\Sigma}_\lambda,\widetilde{\sigma})$ defined by the graph with vertex set $\widetilde{V}_\lambda := \{\widetilde{v}\in\widetilde{V} : \lambda({}_0[\widetilde{v}])>0\}$ and edges accordingly. Since λ is mixing, $(\widetilde{\Sigma}_\lambda,\widetilde{\sigma})$ is topologically mixing (its graph is aperiodic).

Now we prove the theorem. Since the set of periodic points of f is in bijection with the set of periodic points of its natural extension $\widehat{f}$ and $\pi\circ\theta$ is injective, it is enough to estimate $\mathrm{Per}_n(\widetilde{\sigma}\restriction_{\widetilde{\Sigma}_\lambda})$. We consider two cases. In the first, assume that λ is a measure of maximal entropy for $(\widetilde{\Sigma}_\lambda,\widetilde{\sigma})$. By [**Buz20**, Lemma 6.2],

$$\lim_{n\to\infty} e^{-hn}\,\mathrm{Per}_n(\widetilde{\sigma}\restriction_{\widetilde{\Sigma}_\lambda}) \geq 1$$

and the result follows. Now assume that λ is not a measure of maximal entropy. Since $h_{\rm top}(\widetilde{\Sigma}_\lambda)=\sup\{h_{\rm top}(\widetilde{\sigma}\restriction_{\overline{\Sigma}})\}$ where $\overline{\Sigma}$ runs over all subshifts of $\widetilde{\Sigma}_\lambda$ with finitely many states, we can fix one such subshift with $h_{\rm top}(\widetilde{\sigma}\restriction_{\overline{\Sigma}})>h$. The pair $(\overline{\Sigma},\widetilde{\sigma})$ might not be topologically mixing, but since $\overline{\Sigma}$ is defined by a finite graph inside an aperiodic (in particular transitive) graph, we can add to this finite graph finitely many closed paths with coprime lengths and get a new finite graph that is aperiodic and whose associated TMS, still denoted by $\overline{\Sigma}$, satisfies $h_{\rm top}(\widetilde{\sigma}\restriction_{\overline{\Sigma}})=h'>h$. By a standard result for TMS with finitely many states (see e.g. [**Kit98**, Observation 1.4.3]), we have

$$\lim_{n\to\infty} e^{-h'n}\,\mathrm{Per}_n(\widetilde{\sigma}\restriction_{\overline{\Sigma}}) \geq 1$$

and then we obtain a result stronger than the one stated in the theorem. □

CHAPTER 10

Flows with positive speed

In this section we recast some definitions and results of [**LS19**], which will allow us to apply Theorem 1.8 to code nonuniformly hyperbolic measures for flows. We begin defining the flow counterpart of a TMS. Let (Σ, σ) be a TMS and $r : \Sigma \to \mathbb{R}^+$ a Hölder continuous map bounded away from zero and infinity. Let r_n denote the nth Birkhoff sum of r with respect to σ.

TOPOLOGICAL MARKOV FLOW: The *topological Markov flow* with roof function r and base map $\sigma : \Sigma \to \Sigma$ is the flow $\sigma_r : \Sigma_r \to \Sigma_r$ where

$$\Sigma_r := \{(\underline{v}, t) : \underline{v} \in \Sigma, 0 \leq t < r(\underline{v})\},\ \sigma_r^\tau(\underline{v}, t) = (\sigma^n(\underline{v}), t + \tau - r_n(\underline{v}))$$

for the unique $n \in \mathbb{Z}$ s.t. $0 \leq t + \tau - r_n(\underline{v}) < r(\sigma^n(\underline{v}))$.

Informally, σ_r increases the t coordinate at unit speed subject to the identifications $(\underline{v}, r(\underline{v})) \sim (\sigma(\underline{v}), 0)$. The cocycle identity guarantees that $\sigma_r^{\tau_1+\tau_2} = \sigma_r^{\tau_1} \circ \sigma_r^{\tau_2}$. There is a natural metric d_r on Σ_r, called the *Bowen-Walters metric*, s.t. σ_r is a continuous flow [**BW72**]. Moreover, $\exists C > 0, 0 < \kappa < 1$ s.t. $d_r(\sigma_r^\tau(\omega_1), \sigma_r^\tau(\omega_2)) \leq C d_r(\omega_1, \omega_2)^\kappa$ for all $|\tau| < 1$ and $\omega_1, \omega_2 \in \Sigma_r$, see [**LS19**, Lemma 5.8(3)]. Accordingly, we define the *recurrent set* of Σ_r by $\Sigma_r^\# := \{(\underline{v}, t) \in \Sigma_r : \underline{v} \in \Sigma^\#\}$.

Let N be a closed C^∞ Riemannian manifold, let X be a $C^{1+\beta}$ vector field on N s.t. $X_x \neq 0$ for all $x \in N$ and $g = \{g^t\}_{t\in\mathbb{R}} : N \to N$ be the flow defined by X. Given $\chi > 0$, a g-invariant probability measure μ on N is called *χ-hyperbolic* if all of its Lyapunov exponents are greater than χ in absolute value, except for the zero exponent in the flow direction. Theorem 1.8 implies the following result.

THEOREM 10.1. *Let g be as above and $\chi > 0$. If μ is χ-hyperbolic, then there is a topological Markov flow (Σ_r, σ_r) and a Hölder continuous map $\pi_r : \Sigma_r \to N$ s.t.:*

(1) *$r : \Sigma \to \mathbb{R}^+$ is Hölder continuous and bounded away from zero and infinity.*
(2) *$\pi_r \circ \sigma_r^t = g^t \circ \pi_r$ for all $t \in \mathbb{R}$.*
(3) *$\pi_r[\Sigma_r^\#]$ has full measure with respect to μ.*
(4) *Every $x \in N$ has finitely many pre-images in $\Sigma_r^\#$.*

To prove the above theorem, we recall some facts from [**LS19**].

CANONICAL TRANSVERSE BALL AND FLOW BOX: Given $r > 0$, the *canonical transverse ball* at x of radius r is $S_r(x) := \{\exp_x(v) : v \in T_xN, v \perp X_x, \|v\| \leq r\}$, and the *canonical flow box* of $S_r(x)$ is $\mathrm{FB}_r(x) := \{g^t(y) : y \in S_r(x), |t| \leq r\}$.

The following lemmas are the higher dimensional counterparts of [**LS19**, Lemmas 2.2 and 2.3], and their proofs are canonical.

LEMMA 10.2. *There are constants $r_0, \mathfrak{d} \in (0,1)$ which only depend on g s.t. for every $x \in N$ the flow box $\mathrm{FB}_{r_0}(x)$ contains $B(x, \mathfrak{d})$, and the map $(y,t) \mapsto g^t(y)$ is a diffeomorphism from $S_{r_0}(x) \times [-r_0, r_0]$ onto $\mathrm{FB}_{r_0}(x)$.*

LEMMA 10.3. *There is a constant $\mathfrak{H} > 1$ which only depends on g s.t. for every $x \in N$ the maps $\mathfrak{t}_x : \mathrm{FB}_{r_0}(x) \to [-r_0, r_0]$ and $\mathfrak{q}_x : \mathrm{FB}_{r_0}(x) \to S_{r_0}(x)$ defined by $y = \varphi^{\mathfrak{t}_x(y)}[\mathfrak{q}_x(y)]$ are well-defined maps with* $\mathrm{Lip}(\mathfrak{t}_x), \mathrm{Lip}(\mathfrak{q}_x), \|\mathfrak{t}_x\|_{C^{1+\beta}}, \|\mathfrak{q}_x\|_{C^{1+\beta}} \leq \mathfrak{H}$.

STANDARD POINCARÉ SECTION: A *standard Poincaré section* for g is a set $M = \biguplus_{i=1}^n S_r(x_i)$ satisfying the following conditions:

- For every $x \in N$ the set $\{t > 0 : g^t(x) \in M\}$ is a sequence tending to $+\infty$ and the set $\{t < 0 : g^t(x) \in M\}$ is a sequence tending to $-\infty$.
- The *roof function* $R : M \to (0, \infty)$, defined by $R(x) := \min\{t > 0 : g^t(x) \in M\}$, is bounded away from zero and infinity.

POINCARÉ MAP: The *Poincaré map* of a standard Poincaré section M is the map $f : M \to M$ defined by $f(x) := g^{R(x)}(x)$.

Above, r is called the radius of M. Let ∂M denote the relative boundary of M, equal to the disjoint union of the sets $\partial S_r(x_i) := \{\exp_{x_i}(v) : v \in T_{x_i}N, v \perp X_{x_i}, \|v\| = r\}$. The set ∂M introduces discontinuities for f.

SINGULAR AND REGULAR SETS: The *singular set* of f is

$$\mathscr{S} := \left\{ x \in M : \begin{array}{l} x \text{ does } \textit{not} \text{ have a relative neighborhood } V \subset M \setminus \partial M \text{ s.t.} \\ V \text{ is diffeomorphic to an open ball, and } f : V \to f(V) \text{ and} \\ f^{-1} : V \to f^{-1}(V) \text{ are diffeomorphisms} \end{array} \right\}.$$

The *regular set* is $M' = M \setminus \mathscr{S}$.

LEMMA 10.4. *The maps R, f, f^{-1} are differentiable on M', and there is $\mathfrak{C} > 0$ only depending on g s.t. $\|f \restriction_U \|_{C^{1+\beta}} < \mathfrak{C}, \|f^{-1} \restriction_U \|_{C^{1+\beta}} < \mathfrak{C}$ for all open and connected $U \subset M'$.*

This is [**LS19**, Lemma 2.5], and the proof is the same. There is a relation between g-invariant and f-invariant measures.

INDUCED MEASURE: Every g-invariant probability measure μ on N induces an f-invariant measure ν on M by the equality

$$\int_N \varphi d\mu = \frac{1}{\int_M R d\nu} \int_M \left(\int_0^{R(x)} \varphi[g^t(x)]dt \right) d\nu(x), \text{ for all } \varphi \in L^1(\mu).$$

ADAPTED POINCARÉ SECTION: A standard Poincaré section M is *adapted* for μ if $\lim\limits_{n \to \pm\infty} \frac{1}{n} \log d_M(f^n(x), \mathscr{S}) = 0$ for ν-a.e. $x \in M$.

A fortiori, $\nu(\mathscr{S}) = 0$. Above, d_M is the intrinsic Riemannian distance on M, with the convention that the distance between different connected components of M is infinite. The next result states that any g-invariant probability measure has adapted Poincaré sections.

THEOREM 10.5. *Every g-invariant probability measure μ has adapted Poincaré sections with arbitrarily small radius and roof function.*

This is the higher dimensional version of [**LS19**, Theorem 2.8], whose proof consists on constructing a one-parameter family $\{M_r\}_{r \in [a,b]}$ of standard Poincaré sections s.t. M_r has radius r and then using a Borel-Cantelli argument to show that M_r is adapted to Lebesgue a.e. $r \in [a, b]$. To construct the family of standard Poincaré sections, [**LS19**] fixes a finite set of flows boxes that covers N and then takes a fine discretization of the associated canonical transverse balls. Then they show that it is possible to lift/decrease the points of the discretization in the flow direction and take smaller canonical transverse balls centered at these displaced points so that the balls are pairwise disjoint. This argument uses Lemmas 10.2 and 10.3, which are also true in our case. Hence the same proof applies.

Now we discuss nonuniform hyperbolicity. Consider a g-invariant probability measure μ. Let μ be χ-hyperbolic, let M be an adapted Poincaré section for μ, and let ν be the induced measure on M. The derivative cocycle $df_x^n : T_xM \to T_{f^n(x)}M$ is well-defined ν-a.e. By Lemma 10.4, the functions $\log \|df_x\|, \log \|df_x^{-1}\|$ are bounded, hence the Oseledets theorem applies and so f possesses well-defined Lyapunov exponents ν-a.e. The following lemma proves that ν is hyperbolic. Let $\chi' = \chi \inf R$.

LEMMA 10.6. *If μ is χ-hyperbolic for g then ν is χ'-hyperbolic for f.*

This is the higher dimensional version of [**LS19**, Lemma 2.6], and the proof is the same.

PROOF OF THEOREM 10.1. Let μ be χ-hyperbolic, let M be an adapted Poincaré section for μ given by Theorem 10.5. Let $f : M \to M$ be its Poincaré map and ν be the measure on M induced by μ. By Lemma 10.6, ν is χ'-hyperbolic. Since M is the disjoint union of smooth balls of codimension one, assumptions (A1)–(A4) are satisfied. By Lemma 10.4, assumptions (A5)–(A7) are satisfied. Apply Theorem 1.8 for f and the parameter $\chi' > 0$ to get a triple (Σ, σ, π). Since ν is χ'-hyperbolic, it is carried by $\pi[\Sigma^\#] = \mathrm{NUH}_{\chi'}^\#$. Now proceed as in [**LS19**, pp. 230–232] to construct a coding $\pi_r : \Sigma_r \to N$ satisfying Theorem 10.1. □

Concluding this section, we prove Theorems 1.1 and 1.2. Start recalling the definition of the geometric potential of a flow and its equilibrium measures, see e.g. [**LLS16**]. Let $\chi > 0$. The following definition is the flow version of the nonuniformly hyperbolic locus.

THE NONUNIFORMLY HYPERBOLIC LOCUS NUH_χ: It is defined as the set of points $x \in N$ for which there is a splitting $T_xN = E_x^s \oplus X_x \oplus E_x^u$ s.t.:

(NUH1) Every $v \in E_x^s$ contracts in the future at least $-\chi$ and expands in the past:

$$\limsup_{t\to+\infty} \tfrac{1}{t} \log \|dg^t v\| \le -\chi \quad \text{and} \quad \liminf_{t\to+\infty} \tfrac{1}{t} \log \|dg^{-t} v\| > 0.$$

(NUH2) Every $v \in E_x^u$ contracts in the past at least $-\chi$ and expands in the future:

$$\limsup_{t\to+\infty} \tfrac{1}{t} \log \|dg^{-t} v\| \le -\chi \text{ and } \liminf_{t\to+\infty} \tfrac{1}{t} \log \|dg^{t} v\| > 0.$$

(NUH3) The parameters $s(x) = \sup\limits_{\substack{v\in E_x^s \\ \|v\|=1}} S(x,v)$ and $u(x) = \sup\limits_{\substack{w\in E_x^u \\ \|w\|=1}} U(x,w)$ are finite, where:

$$S(x,v) = \sqrt{2}\left(\sum_{n\ge 0} e^{2n\chi} \|dg^n v\|^2\right)^{1/2},$$

$$U(x,w) = \sqrt{2}\left(\sum_{n\ge 0} e^{2n\chi} \|dg^{-n} w\|^2\right)^{1/2}.$$

HYPERBOLIC LOCUS OF g: The *hyperbolic locus* of g is the set $N_{\rm hyp}$ defined as

$$N_{\rm hyp} := \bigcup_{\chi>0} {\rm NUH}_\chi.$$

GEOMETRIC POTENTIAL OF g [**BR75**]: The *geometric potential* of g is the function $J : N_{\rm hyp} \to \mathbb{R}$ defined by

$$J(x) = -\tfrac{d}{dt}\Big|_{t=0} \log\det\left[dg^t \restriction_{E^u_x}\right] = -\lim_{t\to 0}\tfrac{1}{t}\log\det\left[dg^t \restriction_{E^u_x}\right].$$

Observe that J is bounded, since g is $C^{1+\beta}$.

EQUILIBRIUM MEASURE FOR tJ: An *equilibrium measure* for tJ is a g-invariant probability measure μ supported on $N_{\rm hyp}$ s.t. $P_{\rm top}(tJ) = h_\mu(g^1) + \int tJd\mu$, where

$$P_{\rm top}(tJ) := \sup\left\{h_\mu(g^1) + \int_N tJd\mu : \begin{array}{c}\mu \text{ is } g\text{-invariant probability}\\ \text{measure with } \mu(N_{\rm hyp}) = 1\end{array}\right\}.$$

Since J is bounded, $P_{\rm top}(tJ) < \infty$. The proof of Theorem 1.2 is made exactly as in [**LLS16**], since it establishes the "Bernoulli up to a period" property at the symbolic extension and then projects to the flow. The same reasoning can be made to prove Theorem 1.1, using its symbolic version [**LS19**, Theorem 8.1].

CHAPTER 11

Multidimensional billiards

We refer to the books [**CFS82, CM06**] and the survey [**Szá17**] for general texts on the dynamics of billiards. Let $(N, \langle \cdot, \cdot \rangle)$ be a C^∞ compact manifold, let $\{g^t\}_{t \in \mathbb{R}}$ be its geodesic flow, and let $B_1, \dots, B_k$ be open sets of N s.t. their boundaries $\partial B_1, \dots, \partial B_k$ are C^3 submanifolds of codimension one. The sets $B_1, \dots, B_k$ are called *scatterers*. For each $x \in \partial B_i$, let $n(x) \in T_x N$ be the unitary vector orthogonal to $T_x \partial B_i$ that points outward B_i (points in two boundaries have two normal vectors). If $N_0 = N \setminus \bigcup_{i=1}^k B_i$, then we define the *billiard flow* $\varphi = \{\varphi^t\}_{t \in \mathbb{R}}$ on N with scatterers $B_1, \dots, B_k$ as the restriction of the geodesic flow to the unit tangent bundle $T^1 N$ with the identification $v \sim v - 2\langle n(x), v \rangle n(x)$ for $(x, v) \in T^1 N_0$ s.t. $x \in \bigcup_{i=1}^k \partial B_i$. In other words, we flow under the geodesic flow until we hit one of the scatterers, and when this occurs we reflect the vector according to the law of optics, i.e. the angle of incidence equals the angle of reflection. The flow φ is well-defined in a subset of $T^1 N_0$. Letting $M = \{(x, v) \in T^1 N_0 : x \in \bigcup_{i=1}^k \partial B_i \text{ and } \langle v, n(x) \rangle \geq 0\}$, the *billiard map* f is the map of successive hits to the scatterers, i.e. $f(x, v) = (x', v')$ where (x, v) represents a collision point with a direction and (x', v') represents the next collision point with the new direction. Formally, there are closed sets $\mathscr{S}^+, \mathscr{S}^- \subset M$ s.t. $f : M \setminus \mathscr{S}^+ \to M$ and $f^{-1} : M \setminus \mathscr{S}^- \to M$ are diffeomorphisms onto their images, so that we can regard f as an invertible map with singular set $\mathscr{S} = \mathscr{S}^+ \cup \mathscr{S}^-$. When the second fundamental form of the scatterers is positive (the scatterers are strictly convex), the billiard is called *dispersing*, and when it is nonnegative (the scatterers are convex), then the billiard is called *semi-dispersing*. See Figure 11.1 for examples. The map f has a natural invariant Liouville measure $\mu_{\mathrm{SRB}} = \cos \angle(n(x), v) dx dv$.

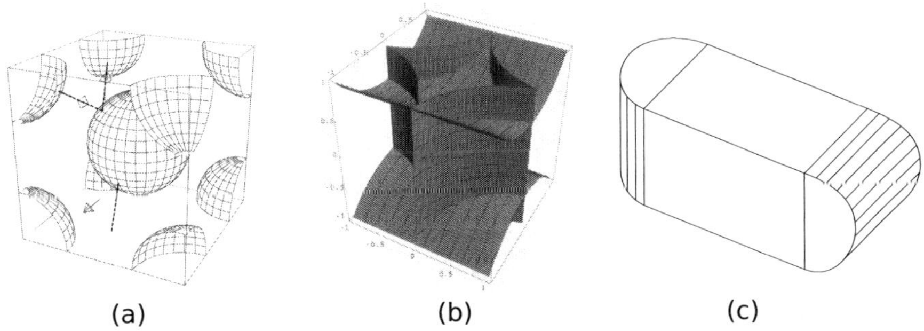

FIGURE 11.1. (a) Dispersing billiard in $\mathbb{T}^3$ (credits to Imre Péter Tóth). (b) Billiard of a hard balls system in $\mathbb{T}^3$ (credits to [**PH00**]). (c) Three-dimensional Bunimovich stadium.

Billiard maps have been studied since the beginning of the twentieth century, one of the reasons being their relation with statistical mechanics. Indeed, the dynamics associated to a mechanical system of m one-dimensional point particles in an interval with heavy walls is a billiard in a subset of $\mathbb{R}^m$. More generally, the famous Lorentz gas and hard balls systems define billiards in some $\mathbb{R}^m$, see [**CFS82**, Chapter 6]. The celebrated *Boltzmann-Sinaĭ hypothesis* states that in billiards of $N \geq 2$ hard balls of unit mass moving on the flat torus $\mathbb{T}^d = \mathbb{R}^d/\mathbb{Z}^d$, $d \geq 2$, the measure $\mu_{\rm SRB}$ is ergodic after the obvious reductions coming from physical invariants [**Sin63**]. The billiards of classical Lorentz gases are dispersing, see Figure 11.1(a). For hard balls systems, the scatterers are cylinders and so the billiards are semidispersing, see Figure 11.1(b).

There is a large literature on billiards, which includes proving the existence of invariant manifolds, nonuniform hyperbolicity, ergodicity, the K property, the Bernoulli property, decay of correlations, etc. The proofs of ergodicity follow Sinaĭ's approach of first obtaining local ergodicity via the so-called *fundamental theorem*. Many authors contributed to this question, among them Sinaĭ, Bunimovich, Chernov, Krámli, Simányi, Szász [**Sin70, BS73, SC87, KSS90**]. While dynamical properties where checked for two-dimensional billiards, most of the above contributions assume that these same properties hold for billiards in higher dimension. For instance, it was commonly assumed that singularity submanifolds have bounded curvature, which turned not to be always true [**BCST03**]. Fortunately, the same authors showed that the theory remains true under a weaker condition called *Lipschitz decomposability*, see [**BCST02**]. They also showed that algebraic semi-dispersing billiards are Lipschitz decomposable (here, algebraic means that the scatterers are defined by algebraic equations). Since billiards coming from hard balls systems are algebraic and $\mu_{\rm SRB}$ is hyperbolic [**SS99**], it was possible to establish the Boltzmann-Sinaĭ ergodic hypothesis, whose contributions include but are not limited to [**KSS91, KSS92, Sim92b, Sim92a, Sim09, Sim13**]. Actually, $\mu_{\rm SRB}$ is Bernoulli [**CH96**].

There is another class of billiards, known as *nowhere dispersing*, where the boundary of scatterers is nowhere convex and yet $\mu_{\rm SRB}$ is hyperbolic. In dimension two, the first examples are due to Bunimovich [**Bun74a, Bun74b, Bun79**], which include the famous Bunimovich stadium. There are some generalizations in higher dimension [**BR97, BR98a, BDM06**], and it is known that $\mu_{\rm SRB}$ is ergodic [**BR98b, BDM08**]. The example considered in [**BDM06, BDM08**] is depicted in Figure 11.1(c), and can be regarded as a three-dimensional version of the Bunimovich stadium.

Ledrappier & Strelcyn have shown that the Pesin entropy formula holds for $\mu_{\rm SRB}$ for a class of systems that includes the billiards mentioned above [**KSLP86**, Part III], therefore in each of these cases $h = h_{\mu_{\rm SRB}}(f) > 0$. The application of Theorem 1.8 for multidimensional billiards deals with the growth rate on the number of periodic orbits. Let $\mathrm{Per}_n(f)$ denote the number of periodic points of period n for f. Up to the authors' knowledge, the only available result about $\mathrm{Per}_n(f)$ in the literature is due to Stoyanov, who proved that for semi-dispersing billiards $\mathrm{Per}_n(f)$ grows at most exponentially fast [**Sto89**]. Our result provides a sufficient condition for the growth rate to be indeed exponential.

THEOREM 11.1. *Let f be a billiard map in any dimension, and assume that $\mu_{\rm SRB}$ is a hyperbolic mixing measure with entropy $h > 0$. Then there exists $C > 0$ s.t.* $\mathrm{Per}_n(f) \geq Ce^{hn}$ *for all n large enough.*

This theorem is a direct consequence of Theorem 9.1 (actually, if $\mu_{\rm SRB}$ is χ-hyperbolic then $\liminf_{n\to\infty} e^{-hn}\,\mathrm{Per}_{n,\chi}(f) \geq 1$). It applies to all examples in Figure 11.1. Note that, the bigger h is, the better is the estimate. The usual way of obtaining sharp estimates is considering measures of maximal entropy, but in the broad setup considered here we do not know if measures of maximal entropy exist nor if they are adapted and hyperbolic. The only situation where these are true is for a class of two-dimensional dispersing billiards [**BD20**]. Using [**LM18, BD20, Buz20**], it follows that $\mathrm{Per}_n(f) \geq Ce^{nh_*}$ for all n large enough, where h_* is the topological entropy of f. It would be interesting to extend this result to higher dimensions.

CHAPTER 12

Viana maps

We finish our applications proving Theorems 1.5 and 1.6. Recall that, for $a_0 \in (1,2)$ s.t. $t=0$ is pre-periodic for the quadratic map $x \mapsto a_0 - t^2$, for $d \geq 16$ and $\alpha > 0$, the associated Viana map is the skew product $f = f_{a_0,d,\alpha} : \mathbb{S}^1 \times \mathbb{R} \to \mathbb{S}^1 \times \mathbb{R}$ defined by $f(\theta,t) = (d\theta, a_0 + \alpha \sin(2\pi\theta) - t^2)$, whose restriction to the annulus $\mathbb{S}^1 \times I_0$ has an attractor. In the sequel, f is considered as this restriction. The map f is C^∞ with a critical set $\mathbb{S}^1 \times \{0\}$ where df is non-invertible and, when $\alpha > 0$ is small, it has a unique invariant probability measure $\mu_{\rm SRB}$ that is absolutely continuous with respect to Lebesgue [**Alv00**, **AV02**], nonuniformly expanding [**Via97**], and mixing [**ALP05**]. All these properties hold for a C^3 neighborhood of f. We claim that assumptions (A1)–(A7) hold for every map in a C^3 neighborhood of f. We also show that, for maps in this neighborhood, every measure with large entropy is adapted. Once these are proved, Theorem 1.5 follows from Theorem 9.1, and Theorem 1.6 follows from Corollary 8.2.

PROPOSITION 12.1. *Every f as above possesses a C^2 neighborhood $\mathscr{U}_f$ s.t. every map in $\mathscr{U}_f$ satisfies assumptions* (A1)–(A7).

PROOF. Start observing that $\mathbb{S}^1 \times \mathbb{R}$ has zero curvature, hence (A1)–(A4) are automatic. It remains to obtain (A5)–(A7). In this proof, we will write $f_0 := f_{a_0,d,\alpha}$ and let f be a map in a C^2 neighborhood of f_0. While we will check (A5) directly, recall from Remark 1.7 that (A6)–(A7) follow from simpler estimates, allowing multiplicative constants and the Lipschitz constants of the derivatives of inverse branches $g : E_x \to g(E_x)$ to be bounded by a multiple of $d(x,\mathscr{S})^{-c}$ for some $c>0$.

The map f_0 is C^2 with $df_0(\theta,t) = \begin{bmatrix} d & 0 \\ * & -2t \end{bmatrix}$, hence it has no discontinuities and its critical/singular set is equal to $\mathscr{S}_{f_0} = \mathbb{S}^1 \times \{0\}$. We fix a C^2 neighborhood $\mathscr{U}$ of f_0 s.t. for all $f \in \mathscr{U}$:

- $\frac{\partial}{\partial t} \det df_x < -2$ for $x = (\theta,t)$.
- $\|df\|_{C^0} < \infty$ and $\min\limits_{x \in \mathbb{S}^1 \times I_0} \|df_x\| > 0$.

Using the first condition, the implicit function theorem implies that the singular set of f is a graph over $\mathbb{S}^1$, i.e. $\mathscr{S}_f = \{(\theta,t_\theta); \theta \in \mathbb{S}^1\}$.

Fix $f \in \mathscr{U}$, and write $\mathscr{S} = \mathscr{S}_f$ and $h(x) = |\det df_x| = \|df_x\| \cdot \|(df_x)^{-1}\|^{-1}$. We start estimating $\|(df_x)^{-1}\|^{-1}$ from below in terms of $d(x,\mathscr{S})$. Fix $x \in \mathbb{S}^1 \times I_0 \backslash \mathscr{S}$. By assumption, $\left|\frac{\partial h}{\partial t}\right| > 2$ and so by the mean value theorem $h(x) = h(\theta,t) - h(\theta,t_\theta) \geq 2|t - t_\theta| \geq 2d(x,\mathscr{S})$, hence

$$\|(df_x)^{-1}\|^{-1} = \tfrac{h(x)}{\|df_x\|} \geq 2\|df\|_{C^0}^{-1} d(x,\mathscr{S}).$$

By the quantitative version of the inverse function theorem, see e.g. [**Sma74**], there is a constant $K = K(f)$ s.t. f is invertible in a ball centered at x with radius

$K\|(df_x)^{-1}\|^{-1} \geq 2K\|df\|_{C^0}^{-1}d(x,\mathscr{S})$. Renaming and reducing K if necessary, we can take $r_1(x) := Kd(x,\mathscr{S})$ with $K < \frac{1}{2}$ and assume that the restriction of f to $B_x := B(x, r_1(x))$ is a diffeomorphism. Now we prove (A5)–(A7).

(A5) If $y \in B_x$ then $d(y,\mathscr{S}) \geq \frac{1}{2}d(x,\mathscr{S})$, hence $\|(df_y)^{-1}\|^{-1} \geq 2\|df\|_{C^0}^{-1}d(y,\mathscr{S}) \geq \|df\|_{C^0}^{-1}d(x,\mathscr{S})$. Therefore $f(B_x)$ contains the ball with center $f(x)$ and radius $\|df\|_{C^0}^{-1}d(x,\mathscr{S})r_1(x) = K\|df\|_{C^0}^{-1}d(x,\mathscr{S})^2$ and so $\mathfrak{r}(x) := \frac{1}{2}K\|df\|_{C^0}^{-1}d(x,\mathscr{S})^2$ satisfies (A5) up to multiplicative constants.

(A6) By hypothesis, $\|df\|_{C^0} < \infty$ and $\min_{y \in \mathbb{S}^1 \times I_0} \|df_y\| > 0$ and so the assumption on df is satisfied up to multiplicative constants. Now let $g : E_x \to g(E_x)$ be the inverse branch of f taking $f(x)$ to x. Let $z \in E_x$ and $g(z) = y \in B_x$. We have $\|dg_z\| = \|(df_y)^{-1}\| \geq \frac{1}{\|df_y\|} \geq \|df\|_{C^0}^{-1}$ and

$$\|dg_z\| = \|(df_y)^{-1}\| = \tfrac{\|df_y\|}{h(y)} \leq \tfrac{\|df\|_{C^0}}{2d(y,\mathscr{S})} \leq \|df\|_{C^0} d(x,\mathscr{S})^{-1},$$

thus proving that dg satisfied (A6) up to multiplicative constants.

(A7) The estimate on df holds because f has bounded C^2 norm. We now estimate the Lipschitz constant of dg. If $z, w \in E_x$ then

$$\begin{aligned}
\|dg_z - dg_w\| &= \|(df_{g(z)})^{-1} - (df_{g(w)})^{-1}\| \\
&\leq \|(df_{g(z)})^{-1}\| \cdot \|(df_{g(w)})^{-1}\| \cdot \|df_{g(z)} - df_{g(w)}\| \\
&\leq \|d^2 f\|_{C^0} \cdot \|(df_{g(z)})^{-1}\| \cdot \|(df_{g(w)})^{-1}\| \cdot d(g(z), g(w)) \\
&\leq \left(\|df\|_{C^0}\|d^2 f\|_{C^0}\|(df_{g(z)})^{-1}\|\|(df_{g(w)})^{-1}\|d(x,\mathscr{S})^{-1}\right) d(z,w) \\
&\leq \left(\|df\|_{C^0}^3\|d^2 f\|_{C^0} d(x,\mathscr{S})^{-3}\right) d(z,w),
\end{aligned}$$

hence we obtain (A7) for g up to multiplicative constants. □

PROPOSITION 12.2. *If f is as above, then every ergodic f-invariant measure μ with $h_\mu(f) \geq h_{\mu_{\rm SRB}}(f)$ is hyperbolic and f-adapted. The same applies to every map that is C^3 close to f.*

PROOF. Similarly to the previous proposition, write $f_0 := f_{a_0,d,\alpha}$ and let f be a map in a C^3 neighborhood of f_0. Given $(\theta_0, t_0) \in \mathbb{S}^1 \times I_0$, we set $(\theta_\ell, t_\ell) := f_0^\ell(\theta_0, t_0)$. Thus

$$(df_0^n)_{(\theta_0,t_0)} = \begin{bmatrix} d^n & 0 \\ b_n(\theta_0,t_0) & c_n(\theta_0,t_0) \end{bmatrix}$$

where $c_n(\theta_0,t_0) = (-2)^n \prod_{\ell=0}^{n-1} t_\ell$ and $|b_n(\theta,t)| \leq nd^n$ for all $n \geq 1$, hence $\lim_{n\to\infty} \frac{1}{n}\log\|(df_0^n)_x\| = \log(d)$ for all $x \in \mathbb{S}^1 \times I_0$. Moreover, if $\varepsilon > 0$ then there is $\delta > 0$ s.t. every f with $d_{C^1}(f, f_0) < \delta$ satisfies

$$\log(d) - \varepsilon \leq \liminf_{n\to\infty} \tfrac{1}{n}\log\|df_x^n\| \leq \limsup_{n\to\infty} \tfrac{1}{n}\log\|df_x^n\| \leq \log(d) + \varepsilon.$$

This implies that if $\lambda_1 \geq \lambda_2$ are the Lyapunov exponents of f with respect to $\mu_{\rm SRB}$, then $\log(d) - \varepsilon \leq \lambda_1 \leq \log(d) + \varepsilon$. By [**Via97**], there is a constant $c = c(f_0) > 0$ s.t. if f is C^3 close to f_0 then $\lambda_2 > c$. In the sequel, we fix $\varepsilon < \frac{c}{2}$ and take f s.t. both estimates on λ_1, λ_2 hold.

Let μ be an ergodic f-invariant probability measure with $h_\mu(f) \geq h_{\mu_{\rm SRB}}(f)$. We first prove that μ is hyperbolic. By the Pesin entropy formula (see [**Liu98**] for a version for endomorphisms), $h_{\mu_{\rm SRB}}(f) = \lambda_1 + \lambda_2$ and by Ruelle inequality we have $h_\mu(f) \leq \overline{\lambda}_1 + \max\{0, \overline{\lambda}_2\}$, where $\overline{\lambda}_1 \geq \overline{\lambda}_2$ are the Lyapunov exponents of f for μ.

Let us prove that $\overline{\lambda}_2 > 0$. By contradiction assume $\overline{\lambda}_2 \leq 0$. Recall that by assumption, $|\lambda_1 - \overline{\lambda}_1| \leq 2\varepsilon$ and so the condition $\overline{\lambda}_1 \geq h_\mu(f) \geq h_{\mu_{\rm SRB}}(f)$ implies that

$$\overline{\lambda}_1 \geq \lambda_1 + \lambda_2$$

which gives that $0 \geq \lambda_2 - |\lambda_1 - \overline{\lambda}_1| > c - 2\varepsilon > 0$, a contradiction. This proves that μ is hyperbolic and $\overline{\lambda}_2 > 0$. It remains to prove that it is f-adapted. Since f is C^2, there is a constant $C > 0$ s.t. $\det df_x \leq C d(x, \mathscr{S})$ and so it is enough to show that $\log(\det df_x) \in L^1(\mu)$. This is easy: first observe that $\log^+(\det df_x)$ is uniformly bounded, then $\log^+(\det df_x) \in L^1(\mu)$ so we can apply Kingman theorem to $\log(\det df_x)$. Recall that $\lim_{n\to\infty} \frac{1}{n} \log(\det df^n_x) = \overline{\lambda}_1 + \overline{\lambda}_2$, μ almost everywhere. Kingman theorem implies that $\int \lim_{n\to\infty} \frac{1}{n} \log(\det df^n_x) d\mu = \int \log(\det df_x) d\mu$, hence $\int \log(\det df_x) d\mu \in (0, \infty)$. The proof is complete. □

APPENDIX A

The graph transform method

In this appendix, we explain how to define and construct the graph transforms induced by a map that is $C^{1+\delta}$ close to a hyperbolic matrix, for some $\delta > 0$. The results proved here are used in Section 4.4. We start recalling some definitions introduced in Section 2.1. Let $\|\cdot\|$ be the euclidean norm of $\mathbb{R}^n$, induced by the canonical inner product $\langle\cdot,\cdot\rangle_{\mathbb{R}^n}$ of $\mathbb{R}^n$. Given a linear transformation $T : \mathbb{R}^n \to \mathbb{R}^n$, let

$$\|T\| = \sup_{v\in\mathbb{R}^n\setminus\{0\}} \tfrac{\|Tv\|}{\|v\|}.$$

Given an open bounded set $U \subset \mathbb{R}^n$ and $h : U \to \mathbb{R}^m$, let $\|h\|_{C^0} := \sup_{x\in U} \|h(x)\|$ denote the C^0 norm of h. When the domain U is not clear in the context, we will write $\|h\|_{C^0(U)}$. For $0 < \delta \leq 1$, let $\mathrm{Hol}_\delta(h) := \sup \frac{\|h(x)-h(y)\|}{\|x-y\|^\delta}$ where the supremum ranges over distinct elements $x, y \in U$. We call h a *contraction* if $\mathrm{Lip}(h) < 1$, and an *expansion* if h is invertible and $\mathrm{Lip}(h^{-1}) < 1$. When h is a linear transformation, then $\|h\| = \mathrm{Lip}(h)$. We also define $\|h\|_{C^\delta} = \|h\|_{C^0} + \mathrm{Hol}_\delta(h)$. If h is differentiable then $\mathrm{Lip}(h) = \|dh\|_{C^0}$. Let $\|h\|_{C^1} := \|h\|_{C^0} + \|dh\|_{C^0} = \|h\|_{C^0} + \mathrm{Lip}(h)$ denote its C^1 norm, and for $0 < \delta \leq 1$ let

$$\|h\|_{C^{1+\delta}} := \|h\|_{C^1} + \mathrm{Hol}_\delta(dh) = \|h\|_{C^0} + \|dh\|_{C^\delta} = \|h\|_{C^0} + \|dh\|_{C^0} + \mathrm{Hol}_\delta(dh)$$

denote its $C^{1+\delta}$ norm. Below we list some basic properties of these norms.

LEMMA A.1. *The following holds:*

(1) *If φ, ψ are maps s.t. $\varphi \circ \psi$ is well-defined, then*

$$\begin{cases}\mathrm{Hol}_\delta(\varphi\circ\psi) \leq \mathrm{Lip}(\varphi)\mathrm{Hol}_\delta(\psi)\\[2ex] \mathrm{Hol}_\delta(\varphi\circ\psi) \leq \mathrm{Hol}_\delta(\varphi)\,\mathrm{Lip}(\psi)^\delta.\end{cases}$$

In particular, if ψ is a contraction then $\|\varphi\circ\psi\|_{C^\delta} \leq \|\varphi\|_{C^\delta}$.

(2) (Perturbation of identity) *If $\|\varphi\|_{C^\delta} < 1$, then $I + \varphi$ is invertible and*

$$\left\|(I+\varphi)^{-1}\right\|_{C^\delta} \leq \tfrac{1}{1-\|\varphi\|_{C^\delta}}.$$

Similarly, if φ is linear and $\|\varphi\| < 1$, then $I + \varphi$ is invertible and

$$\left\|(I+\varphi)^{-1}\right\| \leq \tfrac{1}{1-\|\varphi\|}.$$

Obviously, the euclidean norm is equivalent to any other norm in $\mathbb{R}^n$, but the notions of contraction and expansion differ from the choice of the norm. On the other hand, statements claiming that a norm is small are not very sensitive to the choice of the norm. Given $r > 0$, denote the ball of $\mathbb{R}^n$ with center 0 and radius r by $B^n[r]$. Observe that if $0 \leq d \leq n$ then $B^n[r] \subset B^d[r] \times B^{n-d}[r] \subset B^n[\sqrt{2}r]$.

In this appendix, we fix some parameters:

- $m > 0$, the dimension of the euclidean space.
- $0 \leq d \leq m$, the dimension of one of the invariant subspaces. The other invariant subspace has dimension $m - d$.
- $\delta \in (0, 1)$, the regularity of derivatives.
- $\chi > 0$, that controls the minimal contraction and expansion.
- $\varepsilon > 0$, the parameter of approximation.
- $r > 0$, the size of the domain where the maps will be defined.

The reader should have in mind that the construction will be made for given parameters m, d, δ, χ, and that ε, r will be chosen to make the calculations work. The order of choice will be as follows: $\varepsilon > 0$ will be arbitrarily small to make a finite number of inequalities to hold, and r will be of the form $\varepsilon^{2/\delta} \times C$ with $C < 1$. In particular, $2r^\delta < \varepsilon$ for small $\varepsilon > 0$, and we will use this inequality many times. Since m will be fixed throughout the appendix, we denote the ball of $\mathbb{R}^m$ with center 0 and radius r simply by $B[r]$.

A.1. Admissible graphs

The introduction of admissible graphs requires fixing some extra parameters:

- $p \leq r/2$, the parameter that control the sizes of unstable graphs.
- $q \leq r/2$, the parameter that control the sizes of stable graphs.

All definitions in this section depend on the choice of these parameters. For simplicity of notation, the first d coordinates in $\mathbb{R}^m$ will represent the unstable subspace.

GRAPHS: Given a $C^{1+\delta}$ function $G : B^d[p] \to \mathbb{R}^{m-d}$, let

$$V^u = \{(v, G(v)) : v \in B^d[p]\}$$

denote the graph of G, and call it a *u-graph*. Similarly, given a $C^{1+\delta}$ function $G : B^{m-d}[q] \to \mathbb{R}^d$, let

$$V^s := \{(G(w), w) : w \in B^{m-d}[q]\}$$

denote the graph of G, and call it an *s-graph*. We call G the *representing function*.

Among the s/u-graphs, we are interested in those that can eventually represent stable/unstable manifolds, and for that we require some extra conditions on the representing functions. We call them *admissible graphs*. The definition requires some conditions on the representing function that depends on the minimum $p \wedge q$, so that we can later relate s-graphs and u-graphs. To simplify the exposition and ease the notation, we will work only with unstable graphs with a given parameter p, and introduce an extra parameter $\eta \leq p$ that, for applications in Section 4.4, will be equal to $\eta = p \wedge q$, so that the results obtained in this and next section work for both s-admissible and u-admissible graphs. We will also just use the terminology graph and not mention that it is u or s graph. From now on, fix $\eta \leq p \leq r/2$.

ADMISSIBLE GRAPH: A graph $V = \{(v, G(v)) : v \in B^d[p]\}$ is called an *admissible graph* if its representing function G satisfies the following conditions:

(AM1) $\|G(0)\| \leq 10^{-3}\eta$.
(AM2) $\|(dG)_0\| \leq \frac{1}{2}\eta^\delta$.
(AM3) $\|dG\|_{C^\delta} = \|dG\|_{C^0} + \mathrm{Hol}_\delta(dG) \leq \frac{1}{2}$.

The estimate on the representing function is linear in η, while the estimates on its derivative is of the order of η^δ. This distinction will be clear when we start establishing estimates in the next results. We start with a basic lemma.

LEMMA A.2. *If G is the representing function of an admissible graph, then:*

(1) $\|dG\|_{C^0} < \varepsilon$ *where the norm is taken in* $B^d[p]$.

(2) $\|G\|_{C^0(B^d[t])} < 10^{-2}t$, *for every* $t \in [\eta, p]$.

PROOF.

(1) By (AM2)–(AM3), if $v \in B^d[p]$ then

$$\|(dG)_v\| \leq \|(dG)_0\| + \tfrac{1}{2}\|v\|^\delta \leq \tfrac{1}{2}\eta^\delta + \tfrac{1}{2}p^\delta \leq r^\delta < \varepsilon.$$

(2) By the mean value inequality, the inequality above and (AM1) we obtain that for $v \in B^d[t]$ we have

$$\|G(v)\| \leq \|G(0)\| + \|dG\|_{C^0}\|v\| \leq 10^{-3}\eta + \varepsilon t \leq (10^{-3} + \varepsilon)t < 10^{-2}t,$$

where in the last inequality we require $\varepsilon > 0$ to be small. □

We let $\mathscr{M} = \mathscr{M}^u_{p,\eta}$ denote the set of all admissible graphs as defined above. We define a metric on $\mathscr{M}$, comparing the representing functions. More specifically, for $V_1, V_2 \in \mathscr{M}$ with representing functions G_1, G_2 and for $i \geq 0$, define $d_{C^i}(V_1, V_2) := \|G_1 - G_2\|_{C^i}$ where the norm is taken in $B^d[p]$.

A.2. Graph transforms

Now we introduce a dynamics for which we want to induce an action on admissible graphs. For that, consider a map F that is close to a hyperbolic linear map. We will apply the map F to elements of $\mathscr{M} = \mathscr{M}^u_{p,\eta}$. The idea is to see (a restriction of) each image as an admissible graph in $\widetilde{\mathscr{M}} = \mathscr{M}^u_{\widetilde{p},\widetilde{\eta}}$, where $\widetilde{p}, \widetilde{\eta}$ satisfy $\widetilde{\eta} \leq \widetilde{p}$ and are related to p, η. The relation that the four parameters, $p, \eta, \widetilde{p}, \widetilde{\eta}$, must satisfy are consequences of the contraction/expansion properties required on F. When properly implemented, we will have constructed an operator $\mathscr{F} : \mathscr{M} \to \widetilde{\mathscr{M}}$, called *unstable graph transform* or simply *graph transform*.

We start fixing the four parameters that will define the spaces $\mathscr{M}$ and $\widetilde{\mathscr{M}}$:

- $0 < \eta \leq p$, the parameters defining the domain $\mathscr{M} = \mathscr{M}^u_{p,\eta}$ of $\mathscr{F}$.
- $0 < \widetilde{\eta} \leq \widetilde{p}$, the parameters defining the codomain $\widetilde{\mathscr{M}} = \mathscr{M}^u_{\widetilde{p},\widetilde{\eta}}$ of $\mathscr{F}$.

For $\varepsilon > 0$ small enough, we require the following:

(GT1) $\widetilde{p} \leq e^\varepsilon p$ and $\frac{\widetilde{\eta}}{\eta} = e^{\pm\varepsilon}$.

Now, let $F : B[r] \to \mathbb{R}^m$ be a $C^{1+\delta}$ map of the form $F = D + H$ with

$$D = \begin{bmatrix} D_1 & 0 \\ 0 & D_2 \end{bmatrix}$$

s.t. the following conditions hold:

(GT2) $D_1 : \mathbb{R}^d \to \mathbb{R}^d$ and $D_2 : \mathbb{R}^{n-d} \to \mathbb{R}^{n-d}$ are invertible linear maps s.t. $\|D_1^{-1}\| < e^{-\chi}$ and $\|D_2\| < e^{-\chi}$.

(GT3) $H : B[r] \to \mathbb{R}^m$ satisfies:

(a) $\|H(0)\| < \varepsilon\eta$.

(b) $\|dH\|_{C^0(B[t])} < \varepsilon t^\delta$, for every $t \in [\eta, 2p]$.

(c) $\mathrm{Hol}_\delta(dH) < \varepsilon$, i.e. $\|dH_x - dH_y\| < \varepsilon\|x - y\|^\delta$ for distinct $x, y \in B[r]$.

Let us explain conditions (GT1)–(GT3). Conditions (GT2) and (GT3) state that F is close to a hyperbolic matrix in the $C^{1+\delta}$ norm, where the first d coordinates expand and the last $m-d$ coordinates contract.[1] To explain condition (GT1), consider an admissible graph V with representing function $G : B^d[p] \to \mathbb{R}^{m-d}$. We want to consider $F(V)$ as a new graph with representing function defined on $B^d[\widetilde{p}]$. Since F stretches V horizontally roughly by e^{χ}, if $\varepsilon > 0$ is small enough then this growth is more than e^{ε}. By (GT1) we have $\widetilde{p} \leq e^{\varepsilon}p$, therefore $F(V)$ contains an admissible graph with representing function $\widetilde{G} : B^d[\widetilde{p}] \to \mathbb{R}^{m-d}$. Moreover, by (GT1) the parameters $\eta, \widetilde{\eta}$ are comparable, and by the control on the error term H given by (GT2)–(GT3), we actually obtain that $\widetilde{G}$ satisfies conditions (AM1)–(AM3) for the parameters $\widetilde{p}, \widetilde{\eta}$, thus defining an admissible graph in $\widetilde{\mathscr{M}}$. The attentive reader might have observed the similarity between (GT2)–(GT3) and the statement of Theorem 4.3. It is indeed for $F^{\pm 1}_{\widetilde{x},\widetilde{y}}$ that we will apply the results proved in this appendix.

Observe that, contrary to [**Sar13, BO18**], we do not require any uniform upper bound on $\|D_1\|$ neither lower bound on $\|D_2\|$, so that the results can be applied to the greater generality considered in the main text. Indeed, classical works on the graph transform are set up similar to what we do, see e.g. [**KSLP86, KH95**]. We start defining $\mathscr{F}$ properly.

GRAPH TRANSFORM $\mathscr{F}$: The *graph transform* is the map $\mathscr{F} : \mathscr{M} \to \widetilde{\mathscr{M}}$ that sends an admissible graph $V \in \mathscr{M}$ with representing function G to the unique admissible graph $\widetilde{V}$ with representing function $\widetilde{G} : B^d[\widetilde{p}] \to \mathbb{R}^{m-d}$ s.t. $\{(\widetilde{v}, \widetilde{G}(\widetilde{v})) : \widetilde{v} \in B^d[\widetilde{p}]\} \subset F(V)$.

We already claim that $\widetilde{V} \in \widetilde{\mathscr{M}}$, but this requires checking properties (AM1)–(AM3), that we will do in the sequel. From now on write $H = (h_1, h_2)$, so that

$$F(v, w) = (D_1 v + h_1(v, w), D_2 + h_2(v, w)).$$

A.2.1. Proof that $F(V)$ restricts to a graph. We start showing that $F(V)$ can be restricted to a graph. This means that, in the proper domain, the first coordinate of $F(V)$ is injective.

LEMMA A.3. *The following holds for all $\varepsilon > 0$ small enough. The map $\Psi : B^d[p] \to \mathbb{R}^d$ defined by*

$$\Psi(v) = D_1 v + h_1(v, G(v)) \tag{A.1}$$

is injective, its image $\Psi(B^d[p])$ contains $B^d[e^{\chi - \sqrt{\varepsilon}}p]$, and its inverse $\Phi : \Psi(B^d[p]) \to B^d[p]$ satisfies the following:

(1) $\|d\Phi\|_{C^0} < e^{-\chi+\varepsilon}$.
(2) $\|\Phi(0)\| < 2\varepsilon\eta e^{-\chi+\varepsilon} < 2\varepsilon\eta$.
(3) $\|d\Phi\|_{C^\delta} < e^{-\chi+3\varepsilon}$.

PROOF. This lemma is a sole consequence of the expansion property of D_1. We start proving that Ψ is injective. By Lemma A.2(2), if $v \in B^d[p]$ then $\|G(v)\| < 10^{-2}p$, and so

$$\left\| \begin{bmatrix} v \\ G(v) \end{bmatrix} \right\| < 2p.$$

[1] Recall that we have inverted the order of D_1, D_2 in comparison to the main text.

By (GT3), we know that $\|dh_1\|_{C^0(B[2p])} \le \varepsilon(2p)^\delta < \varepsilon^2$. Given $v_1, v_2 \in B^d[p]$, we have

$$\|(v_1, G(v_1)) - (v_2, G(v_2))\| = \|(v_1 - v_2, G(v_1) - G(v_2))\| \le (1+\varepsilon)\|v_1 - v_2\|,$$

since $\|G(v_1) - G(v_2)\| \le \|dG\|_{C^0}\|v_1 - v_2\| \le \varepsilon\|v_1 - v_2\|$. Now assume that $\Psi(v_1) = \Psi(v_2)$, i.e.

$$D_1(v_1 - v_2) = h_1(v_2, G(v_2)) - h_1(v_1, G(v_1)).$$

We have $\|D_1(v_1 - v_2)\| \ge e^\chi\|v_1 - v_2\|$, while the norm of the right hand side of the above equality is at most

$$\|dh_1\|_{C^0(B[2p])}\|(v_1, G(v_1)) - (v_2, G(v_2))\| \le \varepsilon^2(1+\varepsilon)\|v_1 - v_2\|,$$

which implies that $v_1 = v_2$, hence Ψ is injective. Let $\Phi : \Psi(B^d[p]) \to B^d[p]$ be the inverse of Ψ.

Before continuing, let $L(v)$ be the derivative of the map $v \mapsto h_1(v, G(v))$, i.e.

$$L(v) = (dh_1)_{(v,G(v))}\begin{bmatrix} I \\ dG_v \end{bmatrix}$$

where I is the $d \times d$ identity matrix. By Lemma A.2(1), we have

$$\left\|\begin{bmatrix} I \\ (dG)_v \end{bmatrix}\right\| < 1 + \varepsilon$$

for all $v \in B^d[p]$, hence $\|L\|_{C^0} < 2\|dh_1\|_{C^0(B[2p])} < 2\varepsilon^2$.

Now we show that the image of Ψ contains $B^d[e^{\chi-\sqrt{\varepsilon}}p]$. We prove this using a fixed point theorem. Start observing that if $v \in B^d[p]$ then

$$\|h_1(v, G(v))\| \le \|h_1(0,0)\| + \|dh_1\|_{C^0(B[2p])}\left\|\begin{bmatrix} v \\ G(v) \end{bmatrix}\right\| < \varepsilon\eta + 2\varepsilon^2 p < 2\varepsilon p.$$

If $\widetilde{v} \in B^d[e^{\chi-\sqrt{\varepsilon}}p]$, then $\Psi(v) = \widetilde{v}$ iff v is a fixed point of the map $T : v \in B^d[p] \mapsto D_1^{-1}[\widetilde{v} - h_1(v, G(v))]$. We have

$$\|Tv\| \le \left\|D_1^{-1}\right\|(\|\widetilde{v}\| + \|h_1(v, G(v))\|) < e^{-\chi}\left(e^{\chi-\sqrt{\varepsilon}}p + 2\varepsilon p\right) < \left(e^{-\sqrt{\varepsilon}} + 2\varepsilon\right)p < p,$$

where in the last passage we used that $e^{-\sqrt{\varepsilon}} + 2\varepsilon < 1 - \sqrt{\varepsilon} + 3\varepsilon < 1$ if $\varepsilon > 0$ is small enough. Hence $T : B^d[p] \to B^d[p]$ and, since $\|(dT)_v\| = \left\|D_1^{-1}L(v)\right\| \le \left\|D_1^{-1}\right\|\|L\|_{C^0} < 2\varepsilon^2$, the fixed point theorem implies that T has a unique fixed point. Since $\widetilde{v} \in B^d[e^{\chi-\sqrt{\varepsilon}}p]$ is arbitrary, we get that $\Psi(B^d[p])$ contains $B^d[e^{\chi-\sqrt{\varepsilon}}p]$.

The next step is to prove (1)–(3). We have $(d\Phi)_{\widetilde{v}} = [(d\Psi)_v]^{-1}$ where $v = \Phi(\widetilde{v})$. Letting $\widetilde{L} = L \circ \Phi$, we have $(d\Psi)_v = D_1 + \widetilde{L}(\widetilde{v}) = [I + \widetilde{L}(\widetilde{v})D_1^{-1}]D_1$ and so

$$(d\Phi)_{\widetilde{v}} = D_1^{-1}[I + \widetilde{L}(\widetilde{v})D_1^{-1}]^{-1}.$$

We consider $I + \widetilde{L}(\widetilde{v})D_1^{-1}$ as a perturbation of the identity. We have $\left\|\widetilde{L}D_1^{-1}\right\|_{C^0} = \left\|L\Phi D_1^{-1}\right\|_{C^0} \le \|L\|_{C^0} < 2\varepsilon^2$. For a fixed $\widetilde{v}$, apply Lemma A.1(2) to conclude that $\|(d\Phi)_{\widetilde{v}}\| < \frac{e^{-\chi}}{1-2\varepsilon^2} < e^{-\chi+\varepsilon}$, where in the last inequality we used that $\frac{1}{1-2\varepsilon^2} < 1+\varepsilon < e^\varepsilon$ for $\varepsilon > 0$ small enough. Since $\widetilde{v}$ is arbitrary, we obtain (1).

To prove part (2), we start estimating $\|\Psi(0)\|$. Since $\begin{bmatrix} 0 \\ G(0) \end{bmatrix} \in B[\eta]$, conditions (AM1) and (GT3) give that

$$\|\Psi(0)\| = \|h_1(0, G(0))\| \leq \|h_1(0,0)\| + \|dh_1\|_{C^0(B[\eta])} \left\| \begin{bmatrix} 0 \\ G(0) \end{bmatrix} \right\| \leq \varepsilon\eta + \varepsilon\eta^\delta 10^{-3}\eta$$
$$\leq (\varepsilon + 10^{-3}\varepsilon^2)\eta < 2\varepsilon\eta,$$

hence

$$\|\Phi(0)\| = \|\Phi(0) - \Phi(\Psi(0))\| \leq e^{-\chi+\varepsilon}\|\Psi(0)\| < 2\varepsilon\eta e^{-\chi+\varepsilon}.$$

It remains to prove part (3). Again, we use that $d\Phi = D_1^{-1}[I + \widetilde{L}D_1^{-1}]^{-1}$ and consider $I + \widetilde{L}D_1^{-1}$ as a perturbation of the identity. We will estimate the values of $\mathrm{Hol}_\delta(\widetilde{L}D_1^{-1})$, $\left\|\widetilde{L}D_1^{-1}\right\|_{C^\delta}$ and, using this last estimate, apply Lemma A.1(2) to conclude the proof. Since $\widetilde{L}D_1^{-1} = L\Phi D_1^{-1}$ and Φ, D_1^{-1} are contractions, the second estimate of Lemma A.1(1) implies that $\mathrm{Hol}_\delta(\widetilde{L}D_1^{-1}) \leq \mathrm{Hol}_\delta(L)$, so it is enough to estimate $\mathrm{Hol}_\delta(L)$. Recalling that $\|(v_1, G(v_1)) - (v_2, G(v_2))\| \leq (1+\varepsilon)\|v_1 - v_2\|$ for $v_1, v_2 \in B^d[p]$, we have

$$\begin{aligned}
&\|L(v_1) - L(v_2)\| \\
&= \left\| (dh_1)_{(v_1,G(v_1))} \begin{bmatrix} I \\ (dG)_{v_1} \end{bmatrix} - (dh_1)_{(v_2,G(v_2))} \begin{bmatrix} I \\ (dG)_{v_2} \end{bmatrix} \right\| \\
&\leq \left\| (dh_1)_{(v_1,G(v_1))} \begin{bmatrix} I \\ (dG)_{v_1} \end{bmatrix} - (dh_1)_{(v_1,G(v_1))} \begin{bmatrix} I \\ (dG)_{v_2} \end{bmatrix} \right\| \\
&\quad + \left\| (dh_1)_{(v_1,G(v_1))} \begin{bmatrix} I \\ (dG)_{v_2} \end{bmatrix} - (dh_1)_{(v_2,G(v_2))} \begin{bmatrix} I \\ (dG)_{v_2} \end{bmatrix} \right\| \\
&\leq \|dh_1\|_{C^0(B[2p])} \|(dG)_{v_1} - (dG)_{v_2}\| \\
&\quad + \|(dh_1)_{(v_1,G(v_1))} - (dh_1)_{(v_2,G(v_2))}\| \left\| \begin{bmatrix} I \\ (dG)_{v_2} \end{bmatrix} \right\| \\
&\leq \|dh_1\|_{C^0(B[2p])} \mathrm{Hol}_\delta(dG) \|v_1 - v_2\|^\delta + (1+\varepsilon)^{1+\delta} \mathrm{Hol}_\delta(dh_1) \|v_1 - v_2\|^\delta \\
&< \left[\|dh_1\|_{C^0(B[2p])} \mathrm{Hol}_\delta(dG) + (1+3\varepsilon)\mathrm{Hol}_\delta(dh_1) \right] \|v_1 - v_2\|^\delta,
\end{aligned}$$

thus

$$\mathrm{Hol}_\delta(L) \leq \|dh_1\|_{C^0(B[2p])} \mathrm{Hol}_\delta(dG) + (1+3\varepsilon)\mathrm{Hol}_\delta(dh_1) < \tfrac{1}{2}\varepsilon^2 + (1+3\varepsilon)\varepsilon < 2\varepsilon$$

and so $\mathrm{Hol}_\delta(\widetilde{L}D_1^{-1}) < 2\varepsilon$. We already know that $\left\|\widetilde{L}D_1^{-1}\right\|_{C^0} < 2\varepsilon^2$, therefore $\left\|\widetilde{L}D_1^{-1}\right\|_{C^\delta} < 2\varepsilon^2 + 2\varepsilon < 2.5\varepsilon$ for $\varepsilon > 0$ small enough. Since $d\Phi = D_1^{-1}[I + \widetilde{L}D_1^{-1}]^{-1}$, Lemma A.1(2) gives that

$$\|d\Phi\|_{C^\delta} \leq \|D_1^{-1}\| \tfrac{1}{1-2.5\varepsilon} < e^{-\chi} \tfrac{1}{1-2.5\varepsilon} < e^{-\chi+3\varepsilon},$$

where in the last passage we used that $\frac{1}{1-2.5\varepsilon} < 1 + 3\varepsilon < e^{3\varepsilon}$ for small $\varepsilon > 0$. □

Remark A.4. The derivative $(d\Psi)_v = D_1 + \widetilde{L}(\widetilde{v})$ has the form $A + B$, which could be written as $(I + BA^{-1})A$ as we did or as $A(I + A^{-1}B)$ as did in [**BO18**]. The problem with this second factorization is that the inverse map has the form $(I + A^{-1}B)^{-1}A^{-1}$ and so, when estimating its δ-Hölder constant, the term $\mathrm{Lip}(A^{-1})^\delta$ appears and gives an estimate that depends on δ and is weaker than part (3).

By the above lemma, $F(v, G(v))$ can be represented in terms of $\widetilde{v} = \Phi^{-1}(v)$ as

$$(D_1 v + h_1(v, G(v)), D_2 G(v) + h_2(v, G(v))) = (\widetilde{v}, \widetilde{G}(\widetilde{v})),$$

where $\widetilde{G} = [D_2 G(\cdot) + h_2(\cdot, G(\cdot))] \circ \Phi = D_2 G\Phi + h_2(\Phi, G\Phi)$. By hypothesis, we have $\widetilde{p} \leq e^{\varepsilon} p < e^{\chi - \sqrt{\varepsilon}} p$ for $\varepsilon > 0$ small enough. Therefore, if $V \in \mathscr{M}$ is an admissible graph with representing function G, then we can restrict $\widetilde{G}$ to $B^d[\widetilde{p}]$, still denoting it by $\widetilde{G} : B^d[\widetilde{p}] \to \mathbb{R}^{m-d}$, and formally define $\mathscr{F}(V)$ as the graph with representing function $\widetilde{G}$:

$$\mathscr{F}(V) = \left\{ (\widetilde{v}, \widetilde{G}(\widetilde{v})) : v \in B^d[\widetilde{p}] \right\}.$$

A.2.2. Proof that $\mathscr{F}(V)$ is admissible. We let $V \in \mathscr{M}$ with representing function G, and $\mathscr{F}(V)$ with representing function $\widetilde{G}$. In the previous section we explicitly described $\widetilde{G}$. Using this, we will now prove that $\widetilde{G}$ satisfies conditions (AM1)–(AM3) with respect to the parameters $\widetilde{p}, \widetilde{n}$, thus proving that $\mathscr{F}(V) \in \widetilde{\mathscr{M}}$. The proofs rely on the contraction property of D_2.

LEMMA A.5. *$\left\| \widetilde{G}(0) \right\| < e^{-\chi} (\|G(0)\| + \sqrt{\varepsilon}\eta)$, and so $\left\| \widetilde{G}(0) \right\| < 10^{-3} \widetilde{\eta}$.*

PROOF. We have $\widetilde{G}(0) = D_2 G(\Phi(0)) + h_2(\Phi(0), G(\Phi(0)))$. Observe that

$$\|G(\Phi(0))\| \leq \|G(0)\| + \|dG\|_{C^0} \|\Phi(0)\|$$

which, by (AM2) and Lemma A.3(2), satisfies

$$\|G(\Phi(0))\| \leq 10^{-3}\eta + 2\varepsilon^2 \eta = (10^{-3} + 2\varepsilon^2)\eta.$$

In particular, $\|G(\Phi(0))\| < \eta$ and so $\left\| \begin{bmatrix} \Phi(0) \\ G(\Phi(0)) \end{bmatrix} \right\| < 2\eta$. Therefore

$$\begin{aligned}
\left\| \widetilde{G}(0) \right\| &\leq \|D_2 G(\Phi(0))\| + \|h_2(\Phi(0), G(\Phi(0)))\| \\
&\leq \|D_2 G(\Phi(0))\| + \|h_2(0,0)\| + \|dh_2\|_{C^0(B[2\eta])} \left\| \begin{bmatrix} \Phi(0) \\ G(\Phi(0)) \end{bmatrix} \right\| \\
&\leq \|D_2\| (\|G(0)\| + \|dG\|_{C^0} \|\Phi(0)\|) + \|h_2(0,0)\| \\
&\qquad + d \|dh_2\|_{C^0(B[2\eta])} \left\| \begin{bmatrix} \Phi(0) \\ G(\Phi(0)) \end{bmatrix} \right\| \\
&\leq e^{-\chi}(\|G(0)\| + 2\varepsilon^2\eta) + \varepsilon\eta + 2\varepsilon^2\eta \\
&= e^{-\chi} \left(\|G(0)\| + 2\varepsilon^2 \eta + e^{\chi}[\varepsilon\eta + 2\varepsilon^2\eta] \right).
\end{aligned}$$

If $\varepsilon > 0$ is small enough then $2\varepsilon^2\eta + e^{\chi}\left(\varepsilon\eta + 2\varepsilon^2\eta\right) = (2\varepsilon^2 + e^{\chi}[\varepsilon + 2\varepsilon^2])\eta < \sqrt{\varepsilon}\eta$, and so

$$\left\| \widetilde{G}(0) \right\| \leq e^{-\chi} \left(\|G(0)\| + \sqrt{\varepsilon}\eta \right).$$

Using (AM1), in particular we have that

$$\left\| \widetilde{G}(0) \right\| \leq e^{-\chi}(10^{-3}\eta + \sqrt{\varepsilon}\eta) = e^{-\chi}(10^{-3} + \sqrt{\varepsilon})\eta \leq e^{-\chi+\varepsilon}(10^{-3} + \sqrt{\varepsilon})\widetilde{\eta} < 10^{-3}\widetilde{\eta},$$

since $e^{-\chi+\varepsilon}(10^{-3} + \sqrt{\varepsilon}) < 10^{-3}$ if $\varepsilon > 0$ is small enough. □

LEMMA A.6. *$\left\| (d\widetilde{G})_0 \right\| < e^{-2\chi+\varepsilon} \left(\|(dG)_0\| + 2\varepsilon^{\delta}\eta^{\delta} \right)$, and so $\left\| (d\widetilde{G})_0 \right\| < \frac{1}{2}\widetilde{\eta}^{\delta}$.*

PROOF. We will use that $\left\|\begin{bmatrix}\Phi(0)\\ G(\Phi(0))\end{bmatrix}\right\| < 2\eta$, as proved Lemma A.5. Since $\widetilde{G} = [D_2G(\cdot) + h_2(\cdot, G(\cdot))] \circ \Phi$, we have that

$$(d\widetilde{G})_{\widetilde{v}} = \left(D_2(dG)_{\Phi(\widetilde{v})} + (dh_2)_{(\Phi(\widetilde{v}),G(\Phi(\widetilde{v})))}\begin{bmatrix} I \\ (dG)_{\Phi(\widetilde{v})}\end{bmatrix}\right)(d\Phi)_{\widetilde{v}}$$

and so

$$(d\widetilde{G})_0 = \left(D_2(dG)_{\Phi(0)} + (dh_2)_{(\Phi(0),G(\Phi(0)))}\begin{bmatrix} I \\ (dG)_{\Phi(0)}\end{bmatrix}\right)(d\Phi)_0.$$

By (AM3) and Lemma A.3(2),

$$\left\|(dG)_{\Phi(0)}\right\| \le \|(dG)_0\| + \tfrac{1}{2}\|\Phi(0)\|^\delta < \|(dG)_0\| + \tfrac{1}{2}(2\varepsilon\eta)^\delta < \|(dG)_0\| + \varepsilon^\delta\eta^\delta,$$

hence

$$\begin{aligned}
\left\|(d\widetilde{G})_0\right\| &< \left(\|D_2\|\left\|(dG)_{\Phi(0)}\right\| + \|dh_2\|_{C^0(B[2\eta])}\right)\|(d\Phi)_0\| \\
&< \left(e^{-\chi}\left\|(dG)_{\Phi(0)}\right\| + 2\varepsilon\eta^\delta\right)\|(d\Phi)_0\| \\
&= e^{-\chi}\|(d\Phi)_0\|\left(\left\|(dG)_{\Phi(0)}\right\| + 2e^{\chi}\varepsilon\eta^\delta\right) \\
&< e^{-2\chi+\varepsilon}\left(\|(dG)_0\| + \varepsilon^\delta\eta^\delta + 2e^{\chi}\varepsilon\eta^\delta\right) \\
&< e^{-2\chi+\varepsilon}\left(\|(dG)_0\| + 2\varepsilon^\delta\eta^\delta\right)
\end{aligned}$$

where in the last passage we used Lemma A.3(1) and that $2e^{\chi}\varepsilon < \varepsilon^\delta$ for $\varepsilon > 0$ small enough. Using (AM2), in particular we obtain that

$$\begin{aligned}
\left\|(d\widetilde{G})_0\right\| &< e^{-2\chi+\varepsilon}\left(\tfrac{1}{2}\eta^\delta + 2\varepsilon^\delta\eta^\delta\right) = e^{-2\chi+\varepsilon}\left(\tfrac{1}{2} + 2\varepsilon^\delta\right)\eta^\delta \\
&\le e^{-2\chi+2\varepsilon}\left(\tfrac{1}{2} + 2\varepsilon^\delta\right)\widetilde{\eta}^\delta < \tfrac{1}{2}\widetilde{\eta}^\delta,
\end{aligned}$$

since $e^{-2\chi+2\varepsilon}\left(\frac{1}{2} + 2\varepsilon^\delta\right) < \frac{1}{2}$ if $\varepsilon > 0$ is small enough. □

LEMMA A.7. *$\left\|d\widetilde{G}\right\|_{C^\delta} < e^{-\chi}(\|dG\|_{C^\delta} + \sqrt{\varepsilon})$, and so $\left\|d\widetilde{G}\right\|_{C^\delta} < \frac{1}{2}$.*

PROOF. We have

$$(d\widetilde{G})_{\widetilde{v}} = \left(D_2(dG)_{\Phi(\widetilde{v})} + (dh_2)_{(\Phi(\widetilde{v}),G(\Phi(\widetilde{v})))}\begin{bmatrix} I \\ (dG)_{\Phi(\widetilde{v})}\end{bmatrix}\right)(d\Phi)_{\widetilde{v}}.$$

Since $\|d\Phi\|_{C^\delta} < 1$, Lemma A.1(1) implies that

$$\begin{aligned}
\left\|d\widetilde{G}\right\|_{C^\delta} &\le \left\|D_2(dG)_{\Phi(\widetilde{v})} + (dh_2)_{(\Phi(\widetilde{v}),G(\Phi(\widetilde{v})))}\begin{bmatrix} I \\ (dG)_{\Phi(\widetilde{v})}\end{bmatrix}\right\|_{C^\delta} \\
&\le \|D_2\|\|dG\|_{C^\delta} + \left\|(dh_2)_{(\Phi(\widetilde{v}),G(\Phi(\widetilde{v})))}\begin{bmatrix} I \\ (dG)_{\Phi(\widetilde{v})}\end{bmatrix}\right\|_{C^\delta}.
\end{aligned}$$

Call

$$\widetilde{L} = (dh_2)_{(\Phi(\widetilde{v}),G(\Phi(\widetilde{v})))}\begin{bmatrix} I \\ (dG)_{\Phi(\widetilde{v})}\end{bmatrix}.$$

Then $\widetilde{L} = L \circ \Phi$, with $L, \widetilde{L}$ defined as in the proof of Lemma A.3, with h_1 replaced by h_2. Since we have the same estimates for h_1, h_2, we obtain the same estimates for L, i.e. $\|L\|_{C^0} < 2\varepsilon^2$ and $\mathrm{Hol}_\delta(L) < 2\varepsilon$. Since Φ is a contraction, we conclude that $\left\|\widetilde{L}\right\|_{C^\delta} \leq \|L\|_{C^\delta} < 2\varepsilon^2 + 2\varepsilon < 3\varepsilon$ for $\varepsilon > 0$ small enough. Therefore,

$$\left\|d\widetilde{G}\right\|_{C^\delta} \leq e^{-\chi}\|dG\|_{C^\delta} + 3\varepsilon = e^{-\chi}\left(\|dG\|_{C^\delta} + 3\varepsilon e^{\chi}\right) < e^{-\chi}\left(\|dG\|_{C^\delta} + \sqrt{\varepsilon}\right)$$

for $\varepsilon > 0$ small enough. In particular, using (AM3) we get that $\left\|d\widetilde{G}\right\|_{C^\delta} < e^{-\chi}\left(\frac{1}{2} + \sqrt{\varepsilon}\right) < \frac{1}{2}$ when $\varepsilon > 0$ is small enough. □

By Lemmas A.3, A.5, A.6, A.7, we finally obtain that the graph transform $\mathscr{F} : \mathscr{M} \to \widetilde{\mathscr{M}}$ is well-defined.

Remark A.8. The proofs of Lemmas A.5, A.6, A.7 rely solely on our underlying assumptions and on Lemma A.3. Therefore, the obtained estimates hold for $\widetilde{G}$ defined in the larger domain $\Psi(B^d[p])$ and, in particular, $\left\|d\widetilde{G}\right\|_{C^0} < \varepsilon$.

A.3. Estimates on the norm of $\mathscr{F}$

In this section we obtain estimates of $\mathscr{F}$ in the C^0 and C^1 norms.

PROPOSITION A.9. *The following holds for $\varepsilon > 0$ small enough. If $V_1, V_2 \in \mathscr{M}$ then:*

(1) $d_{C^0}(\mathscr{F}(V_1), \mathscr{F}(V_2)) \leq e^{-\chi/2} d_{C^0}(V_1, V_2)$.
(2) $d_{C^1}(\mathscr{F}(V_1), \mathscr{F}(V_2)) \leq e^{-\chi/2}(d_{C^1}(V_1, V_2) + d_{C^0}(V_1, V_2)^\delta)$.

PROOF. Let V_1, V_2 with representing functions G_1, G_2, and let $\widetilde{G}_1, \widetilde{G}_2$ be the representing functions of $\mathscr{F}(V_1), \mathscr{F}(V_2)$. With respect to G_1, G_2, we consider the functions Ψ_1, Ψ_2 as defined in the proof of Lemma A.3, as well as their inverses Φ_1, Φ_2. We also let

$$\Delta_1 = D_2 G_1(\cdot) + h_2(\cdot, G_1(\cdot))$$
$$\Delta_2 = D_2 G_2(\cdot) + h_2(\cdot, G_2(\cdot)),$$

so that $\widetilde{G}_1 = \Delta_1 \Phi_1$ and $\widetilde{G}_2 = \Delta_2 \Phi_2$ (compositions).

(1) We want to prove that $\left\|\widetilde{G}_1 - \widetilde{G}_2\right\|_{C^0} \leq e^{-\chi/2}\|G_1 - G_2\|_{C^0}$. We have

$$\begin{aligned}\left\|\widetilde{G}_1 - \widetilde{G}_2\right\|_{C^0} &\leq \|\Delta_1\Phi_1 - \Delta_1\Phi_2\|_{C^0} + \|\Delta_1\Phi_2 - \Delta_2\Phi_2\|_{C^0} \\ &\leq \|d\Delta_1\|_{C^0}\|\Phi_1 - \Phi_2\|_{C^0} + \|\Delta_1 - \Delta_2\|_{C^0},\end{aligned}$$

so we estimate each of the terms above.

- $\|d\Delta_1\|_{C^0} < 2\varepsilon$: to prove this, note that $d\Delta_1 = D_2 \circ dG_1 + L_1$, where L_1 is a function just like the one considered in Lemma A.3, so in particular $\|L_1\|_{C^0} < 2\varepsilon^2$. Together with Lemma A.2(1), we obtain that

$$\|d\Delta_1\|_{C^0} \leq \|D_2\|\|dG_1\|_{C^0} + \|L_1\|_{C^0} \leq e^{-\chi}\varepsilon + 2\varepsilon^2 < 2\varepsilon.$$

- $\|\Delta_1 - \Delta_2\|_{C^0} \leq (e^{-\chi} + \varepsilon^2)\|G_1 - G_2\|_{C^0}$: to prove this, just observe that

$$\begin{aligned}\|\Delta_1 - \Delta_2\|_{C^0} &= \|D_2(G_1 - G_2) + [h_2(\cdot, G_1(\cdot)) - h_2(\cdot, G_2(\cdot))]\|_{C^0} \\ &\leq \|D_2\|\|G_1 - G_2\|_{C^0} + \|dh_2\|_{C^0(B[2p])}\|G_1 - G_2\|_{C^0} \\ &\leq (e^{-\chi} + \varepsilon^2)\|G_1 - G_2\|_{C^0}.\end{aligned}$$

- $\|\Phi_1 - \Phi_2\|_{C^0} \leq \varepsilon^2 \|G_1 - G_2\|_{C^0}$: start observing that

$$\|\Psi_1 - \Psi_2\|_{C^0} = \|h_1(\cdot, G_1(\cdot)) - h_1(\cdot, G_2(\cdot))\|_{C^0} \leq \varepsilon^2 \|G_1 - G_2\|_{C^0}.$$

It is worth reminding that the norms above are calculated in $B^d[p]$, while the norm of $\Phi_1 - \Phi_2$ is to be calculated in $B^d[e^{\chi - \sqrt{\varepsilon}} p]$. Since $\Psi_1(B^d[p]) \supset B^d[e^{\chi-\sqrt{\varepsilon}}p]$,

$$\begin{aligned}\|\Phi_1 - \Phi_2\|_{C^0} &\leq \|\Phi_1\Psi_1 - \Phi_2\Psi_1\|_{C^0} \\ &= \|\mathrm{Id} - \Phi_2\Psi_1\|_{C^0} \\ &= \|\Phi_2\Psi_2 - \Phi_2\Psi_1\|_{C^0} \\ &\leq \|d\Phi_2\|_{C^0}\|\Psi_2 - \Psi_1\|_{C^0} \\ &\leq \|\Psi_2 - \Psi_1\|_{C^0} \\ &\leq \varepsilon^2\|G_1 - G_2\|_{C^0},\end{aligned}$$

where we used Lemma A.3(1).

Using the three estimates above, we get that

$$\begin{aligned}\left\|\widetilde{G}_1 - \widetilde{G}_2\right\|_{C^0} &\leq 2\varepsilon \cdot \varepsilon^2 \|G_1 - G_2\|_{C^0} + (e^{-\chi} + \varepsilon^2)\|G_1 - G_2\|_{C^0} \\ &= (e^{-\chi} + \varepsilon^2 + 2\varepsilon^3)\|G_1 - G_2\|_{C^0} \\ &\leq (e^{-\chi} + 3\varepsilon^2)\|G_1 - G_2\|_{C^0}\end{aligned}$$

and so $\left\|\widetilde{G}_1 - \widetilde{G}_2\right\|_{C^0} \leq e^{-\chi/2}\|G_1 - G_2\|_{C^0}$ for $\varepsilon > 0$ small enough.

(2) In view of part (1), we need to estimate $\left\|d\widetilde{G}_1 - d\widetilde{G}_2\right\|_{C^0}$ in terms of $\|G_1 - G_2\|_{C^0}$ and $\|G_1 - G_2\|^{\delta}_{C^0}$. We will show that

$$\left\|d\widetilde{G}_1 - d\widetilde{G}_2\right\|_{C^0} \leq \left(e^{-\chi} + 3\varepsilon^2\right)\|dG_1 - dG_2\|_{C^0} + \left(e^{-\chi+\varepsilon} + 3\varepsilon^2\right)\|G_1 - G_2\|^{\delta}_{C^0}.$$

Note that the contribution of $\|G_1 - G_2\|^{\delta}_{C^0}$ is very small. We have $d\widetilde{G}_1 = (d\Delta_1)_{\Phi_1} \circ d\Phi_1$ and $d\widetilde{G}_2 = (d\Delta_2)_{\Phi_2} \circ d\Phi_2$, hence for $\widetilde{v} \in B^d[\widetilde{p}]$:

$$\begin{aligned}\left\|(d\widetilde{G}_1)_{\widetilde{v}} - (d\widetilde{G}_2)_{\widetilde{v}}\right\| &= \left\|(d\Delta_1)_{\Phi_1(\widetilde{v})}(d\Phi_1)_{\widetilde{v}} - (d\Delta_2)_{\Phi_2(\widetilde{v})}(d\Phi_2)_{\widetilde{v}}\right\| \\ &\leq \left\|(d\Delta_1)_{\Phi_1(\widetilde{v})}(d\Phi_1)_{\widetilde{v}} - (d\Delta_1)_{\Phi_1(\widetilde{v})}(d\Phi_2)_{\widetilde{v}}\right\| \\ &\quad + \left\|(d\Delta_1)_{\Phi_1(\widetilde{v})}(d\Phi_2)_{\widetilde{v}} - (d\Delta_1)_{\Phi_2(\widetilde{v})}(d\Phi_2)_{\widetilde{v}}\right\| \\ &\quad + \left\|(d\Delta_1)_{\Phi_2(\widetilde{v})}(d\Phi_2)_{\widetilde{v}} - (d\Delta_2)_{\Phi_2(\widetilde{v})}(d\Phi_2)_{\widetilde{v}}\right\| \\ &\leq \|d\Delta_1\|_{C^0}\|d\Phi_1 - d\Phi_2\|_{C^0} + \mathrm{Hol}_\delta(d\Delta_1)\|\Phi_1 - \Phi_2\|^{\delta}_{C^0}\|d\Phi_2\|_{C^0} \\ &\quad + \|d\Delta_1 - d\Delta_2\|_{C^0}\|d\Phi_2\|_{C^0}.\end{aligned}$$

In the above expression, we need to estimate the values of $\|d\Delta_1 - d\Delta_2\|_{C^0}$, $\mathrm{Hol}_\delta(d\Delta_1)$ and $\|d\Phi_1 - d\Phi_2\|_{C^0}$.

CLAIM 1: $\|d\Delta_1 - d\Delta_2\|_{C^0} \leq (e^{-\chi} + \varepsilon^2)\|dG_1 - dG_2\|_{C^0} + 2\varepsilon\|G_1 - G_2\|^{\delta}_{C^0}$.

PROOF OF CLAIM 1. We have $d\Delta_1 = D_2 \circ dG_1 + L_1$ and $d\Delta_2 = D_2 \circ dG_2 + L_2$ and so

$$\|d\Delta_1 - d\Delta_2\|_{C^0} \leq \|D_2\|\|dG_1 - dG_2\|_{C^0} + \|L_1 - L_2\|. \tag{A.2}$$

We will prove that $\|L_1 - L_2\| \leq \varepsilon^2 \|dG_1 - dG_2\|_{C^0} + 2\varepsilon \|G_1 - G_2\|_{C^0}^{\delta}$. Indeed, since $\left\| \begin{bmatrix} I \\ (dG_2)_v \end{bmatrix} \right\| \leq 1 + \|dG_2\|_{C^0} < 1 + \varepsilon < 2$, we have

$$\begin{aligned}
\|L_1(v) - L_2(v)\| &= \left\| (dh_2)_{(v,G_1(v))} \begin{bmatrix} I \\ (dG_1)_v \end{bmatrix} - (dh_2)_{(v,G_2(v))} \begin{bmatrix} I \\ (dG_2)_v \end{bmatrix} \right\| \\
&\leq \left\| (dh_2)_{(v,G_1(v))} \begin{bmatrix} I \\ (dG_1)_v \end{bmatrix} - (dh_2)_{(v,G_1(v))} \begin{bmatrix} I \\ (dG_2)_v \end{bmatrix} \right\| \\
&\quad + \left\| (dh_2)_{(v,G_1(v))} \begin{bmatrix} I \\ (dG_2)_v \end{bmatrix} - (dh_2)_{(v,G_2(v))} \begin{bmatrix} I \\ (dG_2)_v \end{bmatrix} \right\| \\
&\leq \|dh_2\|_{C^0(B[2p])} \|(dG_1)_v - (dG_2)_v\| \\
&\quad + 2\mathrm{Hol}_\delta(dh_2) \|G_1(v) - G_2(v)\|^{\delta} \\
&\leq \|dh_2\|_{C^0(B[2p])} \|dG_1 - dG_2\|_{C^0} \\
&\quad + 2\mathrm{Hol}_\delta(dh_2) \|G_1 - G_2\|_{C^0}^{\delta}
\end{aligned}$$

and so

$$\|L_1 - L_2\|_{C^0} \leq \varepsilon^2 \|dG_1 - dG_2\|_{C^0} + 2\varepsilon \|G_1 - G_2\|_{C^0}^{\delta}.$$

Using (A.2), $\|d\Delta_1 - d\Delta_2\|_{C^0} \leq (e^{-\chi} + \varepsilon^2) \|dG_1 - dG_2\|_{C^0} + 2\varepsilon \|G_1 - G_2\|_{C^0}^{\delta}$. □

CLAIM 2: $\mathrm{Hol}_\delta(d\Delta_1) < \frac{1}{2} e^{-\chi} + 3\varepsilon < 1$.

PROOF OF CLAIM 2. By Lemma A.1(1),

$$\mathrm{Hol}_\delta(d\Delta_1) \leq \|D_2\| \mathrm{Hol}_\delta(dG_1) + \mathrm{Hol}_\delta(L_1) < \tfrac{1}{2} e^{-\chi} + \mathrm{Hol}_\delta(L_1),$$

so it is enough to prove that $\mathrm{Hol}_\delta(L_1) < 3\varepsilon$. Calculating $\|L_1(v_1) - L_1(v_2)\|$ as in the proof of Claim 1, we get that

$$\begin{aligned}
&\|L_1(v_1) - L_1(v_2)\| \\
&\quad = \left\| (dh_2)_{(v_1,G_1(v_1))} \begin{bmatrix} I \\ (dG_1)_{v_1} \end{bmatrix} - (dh_2)_{(v_2,G_1(v_2))} \begin{bmatrix} I \\ (dG_1)_{v_2} \end{bmatrix} \right\| \\
&\quad \leq \left\| (dh_2)_{(v_1,G_1(v_1))} \begin{bmatrix} I \\ (dG_1)_{v_1} \end{bmatrix} - (dh_2)_{(v_1,G_1(v_1))} \begin{bmatrix} I \\ (dG_1)_{v_2} \end{bmatrix} \right\| \\
&\qquad + \left\| (dh_2)_{(v_1,G_1(v_1))} \begin{bmatrix} I \\ (dG_1)_{v_2} \end{bmatrix} - (dh_2)_{(v_2,G_1(v_2))} \begin{bmatrix} I \\ (dG_1)_{v_2} \end{bmatrix} \right\| \\
&\quad \leq \|dh_2\|_{C^0(B[2p])} \|(dG_1)_{v_1} - (dG_1)_{v_2}\| + 2\mathrm{Hol}_\delta(dh_2) \left\| \begin{bmatrix} v_1 - v_2 \\ G_1(v_1) - G_1(v_2) \end{bmatrix} \right\|^{\delta} \\
&\quad \leq \left(\|dh_2\|_{C^0(B[2p])} \mathrm{Hol}_\delta(dG_1) + 2(1+\varepsilon)^{\delta} \mathrm{Hol}_\delta(dh_2) \right) \|v_1 - v_2\|^{\delta}
\end{aligned}$$

and so $\mathrm{Hol}_\delta(L_1) \leq \frac{\varepsilon^2}{2} + 2(1+\varepsilon)^{\delta} \varepsilon < 3\varepsilon$. □

CLAIM 3: $\|d\Phi_1 - d\Phi_2\|_{C^0} \leq \varepsilon^2 \|dG_1 - dG_2\|_{C^0} + 2\varepsilon \|G_1 - G_2\|_{C^0}^{\delta}$.

PROOF OF CLAIM 3. Fix $\widetilde{v} \in B^d[e^{\chi-\sqrt{\varepsilon}}p]$ and let $v = \Phi(\widetilde{v})$. We have

$$\begin{aligned}
\|(d\Phi_1)_{\widetilde{v}} - (d\Phi_2)_{\widetilde{v}}\| &\le \|\mathrm{Id} - (d\Psi_1)_v(d\Phi_2)_{\widetilde{v}}\| \\
&= \|(d\Psi_2)_v(d\Phi_2)_{\widetilde{v}} - (d\Psi_1)_v(d\Phi_2)_{\widetilde{v}}\| \\
&\le \|(d\Psi_2)_v - (d\Psi_1)_v\|
\end{aligned}$$

because $\|d\Phi_1\|, \|d\Phi_2\| < 1$. We have $(\Psi_2 - \Psi_1)(v) = h_1(v, G_2(v)) - h_1(v, G_1(v))$, thus

$$\begin{aligned}
\|d(\Psi_2 - \Psi_1)_v\| &= \left\| (dh_1)_{(v,G_2(v))} \begin{bmatrix} I \\ (dG_2)_v \end{bmatrix} - (dh_1)_{(v,G_1(v))} \begin{bmatrix} I \\ (dG_1)_v \end{bmatrix} \right\| \\
&\le \left\| (dh_1)_{(v,G_2(v))} \begin{bmatrix} I \\ (dG_2)_v \end{bmatrix} - (dh_1)_{(v,G_2(v))} \begin{bmatrix} I \\ (dG_1)_v \end{bmatrix} \right\| \\
&\quad + \left\| (dh_1)_{(v,G_2(v))} \begin{bmatrix} I \\ (dG_1)_v \end{bmatrix} - (dh_1)_{(v,G_1(v))} \begin{bmatrix} I \\ (dG_1)_v \end{bmatrix} \right\| \\
&\le \|dh_1\|_{C^0(B[2p])}\|(dG_1)_v - (dG_2)_v\| + 2\mathrm{Hol}_\delta(dh_1)\|G_2(v) - G_1(v)\|^\delta \\
&\le \|dh_1\|_{C^0(B[2p])}\|dG_1 - dG_2\|_{C^0} + 2\mathrm{Hol}_\delta(dh_1)\|G_1 - G_2\|_{C^0}^\delta \\
&\le \varepsilon^2\|dG_1 - dG_2\|_{C^0} + 2\varepsilon\|G_1 - G_2\|_{C^0}^\delta,
\end{aligned}$$

concluding the proof of Claim 3. □

Plugging all estimates together, we obtain that for $\varepsilon > 0$ small enough it holds

$$\begin{aligned}
\left\|d\widetilde{G}_1 - d\widetilde{G}_2\right\|_{C^0} &\le 2\varepsilon\left(\varepsilon^2\|dG_1 - dG_2\|_{C^0} + 2\varepsilon\|G_1 - G_2\|_{C^0}^\delta\right) \\
&\quad + e^{-\chi+\varepsilon}\left(\varepsilon^2\|G_1 - G_2\|_{C^0}\right)^\delta \\
&\quad + e^{-\chi+\varepsilon}\left[(e^{-\chi} + \varepsilon^2)\|dG_1 - dG_2\|_{C^0} + 2\varepsilon\|G_1 - G_2\|_{C^0}^\delta\right] \\
&\le \left(e^{-\chi} + \varepsilon^2 + 2\varepsilon^3\right)\|dG_1 - dG_2\|_{C^0} \\
&\quad + \left(e^{-\chi+\varepsilon} + 4\varepsilon^2 + 2\varepsilon\right)\|G_1 - G_2\|_{C^0}^\delta \\
&\le \left(e^{-\chi} + 3\varepsilon^2\right)\|dG_1 - dG_2\|_{C^0} + \left(e^{-\chi+\varepsilon} + 3\varepsilon\right)\|G_1 - G_2\|_{C^0}^\delta.
\end{aligned}$$

Finally, with the estimate of part (1) we conclude that $\left\|\widetilde{G}_1 - \widetilde{G}_2\right\|_{C^1}$ is at most

$$(e^{-\chi}+3\varepsilon^2)\|G_1 - G_2\|_{C^0}+\left(e^{-\chi} + 3\varepsilon^2\right)\|dG_1 - dG_2\|_{C^0}+\left(e^{-\chi+\varepsilon} + 3\varepsilon\right)\|G_1 - G_2\|_{C^0}^\delta$$

which is bounded by $e^{-\chi/2}\left(\|G_1 - G_2\|_{C^1} + \|G_1 - G_2\|_{C^0}^\delta\right)$, proving part (2). □

A.4. Relation between s-graphs and u-graphs

We finish this appendix providing two relations between s-graphs and u-graphs. The first deals with intersections of graphs based at the same point, and the second deals with intersections of images of u-graphs under F with s-graphs. Let $\mathscr{M}^u = \mathscr{M}^u_{p,\eta}$ and $\mathscr{M}^s = \mathscr{M}^s_{q,\eta}$.

LEMMA A.10. *The following holds for $\varepsilon > 0$ small enough. For every $V^s \in \mathscr{M}^s$ and $V^u \in \mathscr{M}^u$ it holds:*

(1) *V^s and V^u intersect at a single point w, and $\|w\| \le 50^{-1}\eta$.*
(2) *$w = w(V^s, V^u)$ is a Lipschitz function of the pair (V^s, V^u), with Lipschitz constant $\frac{2}{1-\varepsilon^2}$.*

PROOF.

(1) Fix V^s, V^u with representing functions G, J. Observe that intersections of V^s and V^u correspond to points (v_1, v_2) that satisfy the system:

$$\begin{cases} v_1 = J(v_2) \\ v_2 = G(v_1). \end{cases}$$

This system, in turn, corresponds to v_1 being a fixed point of $J \circ G$, wherever this composition is well-defined (one direction is direct; for the other, if $(J \circ G)(v_1) = v_1$ then defining $v_2 = G(v_1)$ we have $(v_1, v_2) = (v_1, G(v_1)) = (J(v_2), v_2))$. Hence we are reduced to analysing the fixed points of $J \circ G$.

Since G, J are defined in different domains, we first prove the existence of a fixed point for their composition and later prove that it is unique. For the existence, we reduce their domains to $B^d[\eta]$ and $B^{m-d}[\eta]$. By Lemma A.2(2), we have

$$B^d[\eta] \xrightarrow{G} B^{m-d}[10^{-2}\eta] \subset B^{m-d}[\eta] \xrightarrow{J} B^d[10^{-2}\eta],$$

so the restriction of $J \circ G$ to $B^d[\eta]$ is well-defined. We claim that this restriction is a contraction. Indeed, in the respective domains we have $\|dG\|_{C^0}, \|dJ\|_{C^0} < \varepsilon$ and so $\|d(J \circ G)\|_{C^0} < \varepsilon^2 < 1$. This guarantees the existence of a fixed point $v_1 \in B^d[10^{-2}\eta]$, which guarantees the existence of an intersection point $w = (v_1, v_2) \in V^s \cap V^u$ with $\|w\| < 50^{-1}\eta$.

To prove uniqueness, we first extend G, J to domains on the same size. By the Kirszbraun theorem, we can extend G, J to functions $\widetilde{G} : B^d[p \vee q] \to \mathbb{R}^{m-d}$ and $\widetilde{J} : B^{m-d}[p \vee q] \to \mathbb{R}^d$ so that $\text{Lip}(\widetilde{G}) = \text{Lip}(G) < \varepsilon$ and $\text{Lip}(\widetilde{J}) = \text{Lip}(J) < \varepsilon$. The same calculation performed in Lemma A.2(2) gives that

$$B^d[p \vee q] \xrightarrow{\widetilde{G}} B^{m-d}[10^{-2}(p \vee q)] \subset B^{m-d}[p \vee q] \xrightarrow{\widetilde{J}} B^d[10^{-2}(p \vee q)],$$

and by definition the composition $\widetilde{J} \circ \widetilde{G}$ is a contraction. It therefore has the unique fixed point w.

(2) We start with a simple claim about contractions on metric spaces.

CLAIM: Let (X, d) be a complete metric space, let f_1, f_2 be contractions on X, and let x_1, x_2 be the unique fixed points of f_1, f_2. If $d(f_1, f_2) := \sup\{d(f_1(x), f_2(x)) : x \in X\}$, then

$$d(x_1, x_2) \leq \tfrac{d(f_1, f_2)}{1 - \text{Lip}(f_1)}.$$

PROOF OF THE CLAIM. We have

$$\begin{aligned} d(x_1, x_2) &= d(f_1(x_1), f_2(x_2)) \\ &\leq d(f_1(x_1), f_1(x_2)) + d(f_1(x_2), f_2(x_2)) \\ &\leq \text{Lip}(f_1) d(x_1, x_2) + d(f_1, f_2) \end{aligned}$$

and the claim follows. □

Let $V_1^s, V_2^s \in \mathscr{M}^s$ with representing functions G_1, G_2, and $V_1^u, V_2^u \in \mathscr{M}^u$ with representing functions J_1, J_2. Let $f_1 = J_1 \circ G_1$ and $f_2 = J_2 \circ G_2$, and let w_i be the intersection of V_i^s and V_i^u. By part (1), $w_1 = (v_1, v_2)$ and $w_2 = (t_1, t_2)$ where v_1 is the unique fixed point of f_1 and t_1 is the unique fixed point of f_2. Observing that $\mathrm{Lip}(f_1) < \varepsilon^2$ and that

$$\begin{aligned}\|f_1 - f_2\|_{C^0} &\le \|J_1 \circ G_1 - J_1 \circ G_2\|_{C^0} + \|J_1 \circ G_2 - J_2 \circ G_2\|_{C^0}\\ &\le \mathrm{Lip}(J_1)\|G_1 - G_2\|_{C^0} + \|J_1 - J_2\|_{C^0}\\ &\le \varepsilon\|G_1 - G_2\|_{C^0} + \|J_1 - J_2\|_{C^0}\\ &\le \|G_1 - G_2\|_{C^0} + \|J_1 - J_2\|_{C^0},\end{aligned}$$

the claim implies that $\|v_1 - t_1\| \le \frac{\|G_1-G_2\|_{C^0}+\|J_1-J_2\|_{C^0}}{1-\varepsilon^2}$. Proceeding similarly with v_2, t_2, we conclude that

$$\|w_1 - w_2\| \le \frac{2(\|G_1-G_2\|_{C^0}+\|J_1-J_2\|_{C^0})}{1-\varepsilon^2} = \frac{2(d_{C^0}(V_1^s,V_2^s)+d_{C^0}(V_1^u,V_2^u))}{1-\varepsilon^2}. \qquad \square$$

For the second lemma, let F and $\mathscr{F}$ be as in Section A.2, with $\mathscr{F} : \mathscr{M}^u_{p,\eta} \to \mathscr{M}^u_{\widetilde{p},\widetilde{\eta}}$. We let $\widetilde{q} \ge \widetilde{\eta}$ be another parameter and consider $\mathscr{M}^s_{\widetilde{q},\widetilde{\eta}}$.

LEMMA A.11. *For every $V^u \in \mathscr{M}^u_{p,\eta}$ and $V^s \in \mathscr{M}^s_{\widetilde{q},\widetilde{\eta}}$, the intersection $F(V^u) \cap V^s$ consists of a single point.*

PROOF. If G is the representing function of V^u then

$$F(V^u) = \left\{(\widetilde{v}, \widetilde{G}(\widetilde{v})) : \widetilde{v} \in \Psi(B^d[p])\right\},$$

where $\widetilde{G} : \Psi(B^d[p]) \to \mathbb{R}^{m-d}$ is defined as before by $\widetilde{G} = D_2G\Phi + h_2(\Phi, G\Phi)$. By Remark A.8, $\left\|d\widetilde{G}\right\|_{C^0} < \varepsilon$. Hence we can apply the same proof of Lemma A.10(1) and conclude that $F(V^u) \cap V^s$ consists of a single point. $\square$

Bibliography

[Aar97] Jon Aaronson, *An introduction to infinite ergodic theory*, Mathematical Surveys and Monographs, vol. 50, American Mathematical Society, Providence, RI, 1997, DOI 10.1090/surv/050. MR1450400

[ALP05] José F. Alves, Stefano Luzzatto, and Vilton Pinheiro, *Markov structures and decay of correlations for non-uniformly expanding dynamical systems*, Ann. Inst. H. Poincaré Anal. Non Linéaire **22** (2005), no. 6, 817–839, `https://doi-org.revues.math.u-psud.fr/10.1016/j.anihpc.2004.12.002`. MR2172861

[Alv00] José Ferreira Alves, *SRB measures for non-hyperbolic systems with multidimensional expansion* (English, with English and French summaries), Ann. Sci. École Norm. Sup. (4) **33** (2000), no. 1, 1–32, DOI 10.1016/S0012-9593(00)00101-4. MR1743717

[AV02] José F. Alves and Marcelo Viana, *Statistical stability for robust classes of maps with non-uniform expansion*, Ergodic Theory Dynam. Systems **22** (2002), no. 1, 1–32, DOI 10.1017/S0143385702000019. MR1889563

[AW67] R. L. Adler and B. Weiss, *Entropy, a complete metric invariant for automorphisms of the torus*, Proc. Nat. Acad. Sci. U.S.A. **57** (1967), 1573–1576, DOI 10.1073/pnas.57.6.1573. MR212156

[AW70] Roy L. Adler and Benjamin Weiss, *Similarity of automorphisms of the torus*, Memoirs of the American Mathematical Society, No. 98, American Mathematical Society, Providence, R.I., 1970. MR0257315

[Bab02] Martine Babillot, *On the mixing property for hyperbolic systems*, Israel J. Math. **129** (2002), 61–76, DOI 10.1007/BF02773153. MR1910932

[BCFT18] K. Burns, V. Climenhaga, T. Fisher, and D. J. Thompson, *Unique equilibrium states for geodesic flows in nonpositive curvature*, Geom. Funct. Anal. **28** (2018), no. 5, 1209–1259, DOI 10.1007/s00039-018-0465-8. MR3856792

[BCST02] P. Bálint, N. Chernov, D. Szász, and I. P. Tóth, *Multi-dimensional semi-dispersing billiards: singularities and the fundamental theorem*, Ann. Henri Poincaré **3** (2002), no. 3, 451–482, DOI 10.1007/s00023-002-8624-7. MR1915299

[BCST03] Péter Bálint, Nikolai Chernov, Domokos Szász, and Imre Péter Tóth, *Geometry of multi-dimensional dispersing billiards* (English, with English and French summaries), Astérisque **286** (2003), xviii, 119–150. Geometric methods in dynamics. I. MR2052299

[BD04] Mike Boyle and Tomasz Downarowicz, *The entropy theory of symbolic extensions*, Invent. Math. **156** (2004), no. 1, 119–161, DOI 10.1007/s00222-003-0335-2. MR2047659

[BD20] Viviane Baladi and Mark F. Demers, *On the measure of maximal entropy for finite horizon Sinai billiard maps*, J. Amer. Math. Soc. **33** (2020), no. 2, 381–449, DOI 10.1090/jams/939. MR4073865

[BDM06] Leonid A. Bunimovich and Gianluigi Del Magno, *Semi-focusing billiards: hyperbolicity*, Comm. Math. Phys. **262** (2006), no. 1, 17–32, DOI 10.1007/s00220-005-1473-8. MR2200880

[BDM08] Leonid A. Bunimovich and Gianluigi Del Magno, *Semi-focusing billiards: ergodicity*, Ergodic Theory Dynam. Systems **28** (2008), no. 5, 1377–1417, DOI 10.1017/S0143385707000909. MR2449534

[BKL02] Michael Blank, Gerhard Keller, and Carlangelo Liverani, *Ruelle-Perron-Frobenius spectrum for Anosov maps*, Nonlinearity **15** (2002), no. 6, 1905–1973, DOI 10.1088/0951-7715/15/6/309. MR1938476

[BMW12] K. Burns, H. Masur, and A. Wilkinson, *The Weil-Petersson geodesic flow is ergodic*, Ann. of Math. (2) **175** (2012), no. 2, 835–908, DOI 10.4007/annals.2012.175.2.8. MR2993753

[BO18] Snir Ben Ovadia, *Symbolic dynamics for non-uniformly hyperbolic diffeomorphisms of compact smooth manifolds*, J. Mod. Dyn. **13** (2018), 43–113, DOI 10.3934/jmd.2018013. MR3918259

[BO21] Snir Ben Ovadia, *The set of points with Markovian symbolic dynamics for non-uniformly hyperbolic diffeomorphisms*, Ergodic Theory Dynam. Systems **41** (2021), no. 11, 3244–3269, DOI 10.1017/etds.2020.114. MR4321709

[Bow70] Rufus Bowen, *Markov partitions for Axiom* a *diffeomorphisms*, Amer. J. Math. **92** (1970), 725–747, DOI 10.2307/2373370. MR277003

[Bow73] Rufus Bowen, *Symbolic dynamics for hyperbolic flows*, Amer. J. Math. **95** (1973), 429–460, DOI 10.2307/2373793. MR339281

[Bow75] Rufus Bowen, *Equilibrium states and the ergodic theory of Anosov diffeomorphisms*, Lecture Notes in Mathematics, Vol. 470, Springer-Verlag, Berlin-New York, 1975. MR0442989

[BPVL20] Lucas Backes, Mauricio Poletti, Paulo Varandas, and Yuri Lima, *Simplicity of Lyapunov spectrum for linear cocycles over non-uniformly hyperbolic systems*, Ergodic Theory Dynam. Systems **40** (2020), no. 11, 2947–2969, DOI 10.1017/etds.2019.22. With an appendix by Lima. MR4157470

[BR75] Rufus Bowen and David Ruelle, *The ergodic theory of Axiom A flows*, Invent. Math. **29** (1975), no. 3, 181–202, DOI 10.1007/BF01389848. MR380889

[BR97] Leonid A. Bunimovich and Jan Rehacek, *Nowhere dispersing* 3*D billiards with non-vanishing Lyapunov exponents*, Comm. Math. Phys. **189** (1997), no. 3, 729–757, DOI 10.1007/s002200050227. MR1482937

[BR98a] Leonid A. Bunimovich and Jan Rehacek, *How high-dimensional stadia look like*, Comm. Math. Phys. **197** (1998), no. 2, 277–301, DOI 10.1007/s002200050451. MR1652730

[BR98b] Leonid A. Bunimovich and Jan Rehacek, *On the ergodicity of many-dimensional focusing billiards* (English, with English and French summaries), Ann. Inst. H. Poincaré Phys. Théor. **68** (1998), no. 4, 421–448. Classical and quantum chaos. MR1634311

[Bru95] H. Bruin, *Induced maps, Markov extensions and invariant measures in one-dimensional dynamics*, Comm. Math. Phys. **168** (1995), no. 3, 571–580. MR1328254

[BS03] Jérôme Buzzi and Omri Sarig, *Uniqueness of equilibrium measures for countable Markov shifts and multidimensional piecewise expanding maps*, Ergodic Theory Dynam. Systems **23** (2003), no. 5, 1383–1400, DOI 10.1017/S0143385703000087. MR2018604

[BS73] L. A. Bunimovič and Ja. G. Sinaĭ, *The fundamental theorem of the theory of scattering billiards* (Russian), Mat. Sb. (N.S.) **90(132)** (1973), 415–431, 479. MR0367153

[BSC90] L. A. Bunimovich, Ya. G. Sinaĭ, and N. I. Chernov, *Markov partitions for two-dimensional hyperbolic billiards* (Russian), Uspekhi Mat. Nauk **45** (1990), no. 3(273), 97–134, 221, DOI 10.1070/RM1990v045n03ABEH002355; English transl., Russian Math. Surveys **45** (1990), no. 3, 105–152. MR1071936

[Bun74a] L. A. Bunimovič, *Billiards that are close to scattering billiards* (Russian), Mat. Sb. (N.S.) **94(136)** (1974), 49–73, 159. MR0342677

[Bun74b] L. A. Bunimovič, *The ergodic properties of certain billiards* (Russian), Funkcional. Anal. i Priložen. **8** (1974), no. 3, 73–74. MR0357736

[Bun79] L. A. Bunimovich, *On the ergodic properties of nowhere dispersing billiards*, Comm. Math. Phys. **65** (1979), no. 3, 295–312. MR530154

[Bur11] David Burguet, $\mathcal{C}^2$ *surface diffeomorphisms have symbolic extensions*, Invent. Math. **186** (2011), no. 1, 191–236, DOI 10.1007/s00222-011-0317-8. MR2836054

[Buz20] Jérôme Buzzi, *The degree of Bowen factors and injective codings of diffeomorphisms*, J. Mod. Dyn. **16** (2020), 1–36, DOI 10.3934/jmd.2020001. MR4063921

[Buz97] Jérôme Buzzi, *Intrinsic ergodicity of affine maps in* $[0,1]^d$, Monatsh. Math. **124** (1997), no. 2, 97–118, DOI 10.1007/BF01300614. MR1462857

[Buz98] Jérôme Buzzi, *Ergodicité intrinsèque de produits fibrés d'applications chaotiques unidimensionelles* (French, with English and French summaries), Bull. Soc. Math. France **126** (1998), no. 1, 51–77. MR1651381

[Buz99] Jérôme Buzzi, *Markov extensions for multi-dimensional dynamical systems*, Israel J. Math. **112** (1999), 357–380, DOI 10.1007/BF02773488. MR1714974

[BW72] Rufus Bowen and Peter Walters, *Expansive one-parameter flows*, J. Differential Equations **12** (1972), 180–193, DOI 10.1016/0022-0396(72)90013-7. MR341451

[CFS82] I. P. Cornfeld, S. V. Fomin, and Ya. G. Sinaĭ, *Ergodic theory*, Grundlehren der mathematischen Wissenschaften [Fundamental Principles of Mathematical Sciences], vol. 245, Springer-Verlag, New York, 1982. Translated from the Russian by A. B. Sosinskiĭ, DOI 10.1007/978-1-4615-6927-5. MR832433

[CH96] N. I. Chernov and C. Haskell, *Nonuniformly hyperbolic K-systems are Bernoulli*, Ergodic Theory Dynam. Systems **16** (1996), no. 1, 19–44, DOI 10.1017/S0143385700008695. MR1375125

[Che99] N. Chernov, *Decay of correlations and dispersing billiards*, J. Statist. Phys. **94** (1999), no. 3-4, 513–556, DOI 10.1023/A:1004581304939. MR1675363

[CM06] Nikolai Chernov and Roberto Markarian, *Chaotic billiards*, Mathematical Surveys and Monographs, vol. 127, American Mathematical Society, Providence, RI, 2006, DOI 10.1090/surv/127. MR2229799

[CT22] Benjamin Call and Daniel J. Thompson, *Equilibrium states for self-products of flows and the mixing properties of rank 1 geodesic flows*, J. Lond. Math. Soc. (2) **105** (2022), no. 2, 794–824, DOI 10.1112/jlms.12517. MR4400937

[DM09] Tomasz Downarowicz and Alejandro Maass, *Smooth interval maps have symbolic extensions: the antarctic theorem*, Invent. Math. **176** (2009), no. 3, 617–636, DOI 10.1007/s00222-008-0172-4. MR2501298

[DN05] Tomasz Downarowicz and Sheldon Newhouse, *Symbolic extensions and smooth dynamical systems*, Invent. Math. **160** (2005), no. 3, 453–499, DOI 10.1007/s00222-004-0413-0. MR2178700

[Dow11] Tomasz Downarowicz, *Entropy in dynamical systems*, New Mathematical Monographs, vol. 18, Cambridge University Press, Cambridge, 2011, DOI 10.1017/CBO9780511976155. MR2809170

[Had98] J. Hadamard, *Sur la forme des lignes géodésiques à l'infini et sur les géodésiques des surfaces réglées du second ordre* (French), Bull. Soc. Math. France **26** (1898), 195–216. MR1504321

[Hof78] Franz Hofbauer, *β-shifts have unique maximal measure*, Monatsh. Math. **85** (1978), no. 3, 189–198, DOI 10.1007/BF01534862. MR492180

[Hof79] Franz Hofbauer, *On intrinsic ergodicity of piecewise monotonic transformations with positive entropy*, Israel J. Math. **34** (1979), no. 3, 213–237 (1980), DOI 10.1007/BF02760884. MR570882

[Hof81] Franz Hofbauer, *On intrinsic ergodicity of piecewise monotonic transformations with positive entropy. II*, Israel J. Math. **38** (1981), no. 1-2, 107–115, DOI 10.1007/BF02761854. MR599481

[Hub93] J. H. Hubbard, *Local connectivity of Julia sets and bifurcation loci: three theorems of J.-C. Yoccoz*, Topological methods in modern mathematics (Stony Brook, NY, 1991), Publish or Perish, Houston, TX, 1993, pp. 467–511. MR1215974

[Kat80] A. Katok, *Lyapunov exponents, entropy and periodic orbits for diffeomorphisms*, Inst. Hautes Études Sci. Publ. Math. **51** (1980), 137–173. MR573822

[KH95] Anatole Katok and Boris Hasselblatt, *Introduction to the modern theory of dynamical systems*, Encyclopedia of Mathematics and its Applications, vol. 54, Cambridge University Press, Cambridge, 1995. With a supplementary chapter by Katok and Leonardo Mendoza, DOI 10.1017/CBO9780511809187. MR1326374

[Kit98] Bruce P. Kitchens, *Symbolic dynamics*, Universitext, Springer-Verlag, Berlin, 1998. One-sided, two-sided and countable state Markov shifts, DOI 10.1007/978-3-642-58822-8. MR1484730

[Kni98] Gerhard Knieper, *The uniqueness of the measure of maximal entropy for geodesic flows on rank* 1 *manifolds*, Ann. of Math. (2) **148** (1998), no. 1, 291–314, DOI 10.2307/120995. MR1652924

[KSLP86] Anatole Katok, Jean-Marie Strelcyn, F. Ledrappier, and F. Przytycki, *Invariant manifolds, entropy and billiards; smooth maps with singularities*, Lecture Notes in Mathematics, vol. 1222, Springer-Verlag, Berlin, 1986, DOI 10.1007/BFb0099031. MR872698

[KSS90] A. Krámli, N. Simányi, and D. Szász, *A "transversal" fundamental theorem for semi-dispersing billiards*, Comm. Math. Phys. **129** (1990), no. 3, 535–560. MR1051504

[KSS91] A. Krámli, N. Simányi, and D. Szász, *The K-property of three billiard balls*, Ann. of Math. (2) **133** (1991), no. 1, 37–72, DOI 10.2307/2944325. MR1087345

[KSS92] A. Krámli, N. Simányi, and D. Szász, *The K-property of four billiard balls*, Comm. Math. Phys. **144** (1992), no. 1, 107–148. MR1151248

[KU07] Svetlana Katok and Ilie Ugarcovici, *Symbolic dynamics for the modular surface and beyond*, Bull. Amer. Math. Soc. (N.S.) **44** (2007), no. 1, 87–132, DOI 10.1090/S0273-0979-06-01115-3. MR2265011

[Lim20] Yuri Lima, *Symbolic dynamics for one dimensional maps with nonuniform expansion*, Ann. Inst. H. Poincaré C Anal. Non Linéaire **37** (2020), no. 3, 727–755, DOI 10.1016/j.anihpc.2019.10.001. MR4093615

[Liu98] Pei-Dong Liu, *Pesin's entropy formula for endomorphisms*, Nagoya Math. J. **150** (1998), 197–209, DOI 10.1017/S0027763000025113. MR1633167

[LLS16] François Ledrappier, Yuri Lima, and Omri Sarig, *Ergodic properties of equilibrium measures for smooth three dimensional flows*, Comment. Math. Helv. **91** (2016), no. 1, 65–106, DOI 10.4171/CMH/378. MR3471937

[LM18] Yuri Lima and Carlos Matheus, *Symbolic dynamics for non-uniformly hyperbolic surface maps with discontinuities* (English, with English and French summaries), Ann. Sci. Éc. Norm. Supér. (4) **51** (2018), no. 1, 1–38, DOI 10.24033/asens.2350. MR3764037

[LS19] Yuri Lima and Omri M. Sarig, *Symbolic dynamics for three-dimensional flows with positive topological entropy*, J. Eur. Math. Soc. (JEMS) **21** (2019), no. 1, 199–256, DOI 10.4171/JEMS/834. MR3880208

[Mar04] Roberto Markarian, *Billiards with polynomial decay of correlations*, Ergodic Theory Dynam. Systems **24** (2004), no. 1, 177–197, DOI 10.1017/S0143385703000270. MR2041267

[PH00] H. A. Posch and R. Hirschl, *Simulation of billiards and of hard body fluids*, Hard ball systems and the Lorentz gas, Encyclopaedia Math. Sci., vol. 101, Springer, Berlin, 2000, pp. 279–314, DOI 10.1007/978-3-662-04062-1_11. MR1805333

[Pin11] Vilton Pinheiro, *Expanding measures*, Ann. Inst. H. Poincaré C Anal. Non Linéaire **28** (2011), no. 6, 889–939, DOI 10.1016/j.anihpc.2011.07.001. MR2859932

[PS05] Yakov Pesin and Samuel Senti, *Thermodynamical formalism associated with inducing schemes for one-dimensional maps* (English, with English and Russian summaries), Mosc. Math. J. **5** (2005), no. 3, 669–678, 743–744, DOI 10.17323/1609-4514-2005-5-3-669-678. MR2241816

[PSZ08] Ya. B. Pesin, S. Senti, and K. Zhang, *Lifting measures to inducing schemes*, Ergodic Theory Dynam. Systems **28** (2008), no. 2, 553–574, `https://doi.org/10.1017/S0143385707000806`. MR2408392

[Rat69] M. E. Ratner, *Markov decomposition for an U-flow on a three-dimensional manifold* (Russian), Mat. Zametki **6** (1969), 693–704. MR260977

[Rat73] M. Ratner, *Markov partitions for Anosov flows on n-dimensional manifolds*, Israel J. Math. **15** (1973), 92–114, DOI 10.1007/BF02771776. MR339282

[Roh61] V. A. Rohlin, *Exact endomorphisms of a Lebesgue space* (Russian), Izv. Akad. Nauk SSSR Ser. Mat. **25** (1961), 499–530. MR0143873

[Sar11] Omri M. Sarig, *Bernoulli equilibrium states for surface diffeomorphisms*, J. Mod. Dyn. **5** (2011), no. 3, 593–608, DOI 10.3934/jmd.2011.5.593. MR2854097

[Sar13] Omri M. Sarig, *Symbolic dynamics for surface diffeomorphisms with positive entropy*, J. Amer. Math. Soc. **26** (2013), no. 2, 341–426, DOI 10.1090/S0894-0347-2012-00758-9. MR3011417

[SC87] Ya. G. Sinaĭ and N. I. Chernov, *Ergodic properties of some systems of two-dimensional disks and three-dimensional balls* (Russian), Uspekhi Mat. Nauk **42** (1987), no. 3(255), 153–174, 256. MR896880

[Sim09] Nándor Simányi, *Conditional proof of the Boltzmann-Sinai ergodic hypothesis*, Invent. Math. **177** (2009), no. 2, 381–413, DOI 10.1007/s00222-009-0182-x. MR2511746

[Sim13] Nándor Simányi, *Singularities and non-hyperbolic manifolds do not coincide*, Nonlinearity **26** (2013), no. 6, 1703–1717, DOI 10.1088/0951-7715/26/6/1703. MR3065929

[Sim92a] Nándor Simányi, *The K-property of N billiard balls. I*, Invent. Math. **108** (1992), no. 3, 521–548, DOI 10.1007/BF02100616. MR1163237

[Sim92b] Nándor Simányi, *The K-property of N billiard balls. II. Computation of neutral linear spaces*, Invent. Math. **110** (1992), no. 1, 151–172, DOI 10.1007/BF01231329. MR1181821

[Sin63] Ja. G. Sinaĭ, *On the foundations of the ergodic hypothesis for a dynamical system of statistical mechanics*, Soviet Math. Dokl. **4** (1963), 1818–1822. MR0214727

[Sin68a] Ja. G. Sinaĭ, *Construction of Markov partitionings* (Russian), Funkcional. Anal. i Priložen. **2** (1968), no. 3, 70–80 (Loose errata). MR0250352

[Sin68b] Ja. G. Sinaĭ, *Markov partitions and U-diffeomorphisms* (Russian), Funkcional. Anal. i Priložen. **2** (1968), no. 1, 64–89. MR0233038

[Sin70] Ja. G. Sinaĭ, *Dynamical systems with elastic reflections. Ergodic properties of dispersing billiards* (Russian), Uspehi Mat. Nauk **25** (1970), no. 2 (152), 141–192. MR0274721

[Sma74] D. R. Smart, *Fixed point theorems*, Cambridge Tracts in Mathematics, No. 66, Cambridge University Press, London-New York, 1974. MR0467717

[SS99] Nándor Simányi and Domokos Szász, *Hard ball systems are completely hyperbolic*, Ann. of Math. (2) **149** (1999), no. 1, 35–96, DOI 10.2307/121019. MR1680535

[Sto89] Luchezar Stojanov, *An estimate from above of the number of periodic orbits for semi-dispersed billiards*, Comm. Math. Phys. **124** (1989), no. 2, 217–227. MR1012865

[Szá17] Domokos Szász, *Multidimensional hyperbolic billiards*, Dynamical systems, ergodic theory, and probability: in memory of Kolya Chernov, Contemp. Math., vol. 698, Amer. Math. Soc., Providence, RI, 2017, pp. 201–220, DOI 10.1090/conm/698/14028. MR3716095

[Tak73] Yōichirō Takahashi, *Isomorphisms of β-automorphisms to Markov automorphisms*, Osaka Math. J. **10** (1973), 175–184. MR340552

[Via97] Marcelo Viana, *Multidimensional nonhyperbolic attractors*, Inst. Hautes Études Sci. Publ. Math. **85** (1997), 63–96. MR1471866

[You98] Lai-Sang Young, *Statistical properties of dynamical systems with some hyperbolicity*, Ann. of Math. (2) **147** (1998), no. 3, 585–650, DOI 10.2307/120960. MR1637655

Editorial Information

To be published in the *Memoirs*, a paper must be correct, new, nontrivial, and significant. Further, it must be well written and of interest to a substantial number of mathematicians. Piecemeal results, such as an inconclusive step toward an unproved major theorem or a minor variation on a known result, are in general not acceptable for publication.

Papers appearing in *Memoirs* are generally at least 80 and not more than 200 published pages in length. Papers less than 80 or more than 200 published pages require the approval of the Managing Editor of the Transactions/Memoirs Editorial Board. Published pages are the same size as those generated in the style files provided for $\mathcal{A}_{\mathcal{M}}\mathcal{S}$-LaTeX.

Information on the backlog for this journal can be found on the AMS website starting from `http://www.ams.org/memo`.

A Consent to Publish is required before we can begin processing your paper. After a paper is accepted for publication, the Providence office will send a Consent to Publish and Copyright Agreement to all authors of the paper. By submitting a paper to the *Memoirs*, authors certify that the results have not been submitted to nor are they under consideration for publication by another journal, conference proceedings, or similar publication.

Information for Authors

Memoirs is an author-prepared publication. Once formatted for print and on-line publication, articles will be published as is with the addition of AMS-prepared frontmatter and backmatter. Articles are not copyedited; however, confirmation copy will be sent to the authors.

Initial submission. The AMS uses Centralized Manuscript Processing for initial submissions. Authors should submit a PDF file using the Initial Manuscript Submission form found at `www.ams.org/submission/memo`, or send one copy of the manuscript to the following address: Centralized Manuscript Processing, MEMOIRS OF THE AMS, 201 Charles Street, Providence, RI 02904-2294 USA. If a paper copy is being forwarded to the AMS, indicate that it is for *Memoirs* and include the name of the corresponding author, contact information such as email address or mailing address, and the name of an appropriate Editor to review the paper (see the list of Editors below).

The paper must contain a *descriptive title* and an *abstract* that summarizes the article in language suitable for workers in the general field (algebra, analysis, etc.). The *descriptive title* should be short, but informative; useless or vague phrases such as "some remarks about" or "concerning" should be avoided. The *abstract* should be at least one complete sentence, and at most 300 words. Included with the footnotes to the paper should be the 2020 *Mathematics Subject Classification* representing the primary and secondary subjects of the article. The classifications are accessible from `www.ams.org/msc/`. The Mathematics Subject Classification footnote may be followed by a list of *key words and phrases* describing the subject matter of the article and taken from it. Journal abbreviations used in bibliographies are listed in the latest *Mathematical Reviews* annual index. The series abbreviations are also accessible from `www.ams.org/msnhtml/serials.pdf`. To help in preparing and verifying references, the AMS offers MR Lookup, a Reference Tool for Linking, at `www.ams.org/mrlookup/`.

Electronically prepared manuscripts. The AMS encourages electronically prepared manuscripts, with a strong preference for $\mathcal{A}_{\mathcal{M}}\mathcal{S}$-LaTeX. To this end, the Society has prepared $\mathcal{A}_{\mathcal{M}}\mathcal{S}$-LaTeX author packages for each AMS publication. Author packages include instructions for preparing electronic manuscripts, samples, and a style file that generates the particular design specifications of that publication series.

Authors may retrieve an author package for *Memoirs of the AMS* from `www.ams.org/journals/memo/memoauthorpac.html`. The *AMS Author Handbook* is available in PDF format from the author package link. The author package can also be obtained free of charge by sending email to `tech-support@ams.org` or from the Publication Division,

American Mathematical Society, 201 Charles St., Providence, RI 02904-2294, USA. When requesting an author package, please specify the publication in which your paper will appear. Please be sure to include your complete mailing address.

After acceptance. The source files for the final version of the electronic manuscript should be sent to the Providence office immediately after the paper has been accepted for publication. The author should also submit a PDF of the final version of the paper to the editor, who will forward a copy to the Providence office.

Accepted electronically prepared files can be submitted via the web at `www.ams.org/submit-book-journal/`, sent via FTP, or sent on CD to the Electronic Prepress Department, American Mathematical Society, 201 Charles Street, Providence, RI 02904-2294 USA. TEX source files and graphic files can be transferred over the Internet by FTP to the Internet node `ftp.ams.org` (130.44.1.100). When sending a manuscript electronically via CD, please be sure to include a message indicating that the paper is for the *Memoirs*.

Electronic graphics. Comprehensive instructions on preparing graphics are available at `www.ams.org/authors/journals.html`. A few of the major requirements are given here.

Submit files for graphics as EPS (Encapsulated PostScript) files. This includes graphics originated via a graphics application as well as scanned photographs or other computer-generated images. If this is not possible, TIFF files are acceptable as long as they can be opened in Adobe Photoshop or Illustrator.

Authors using graphics packages for the creation of electronic art should also avoid the use of any lines thinner than 0.5 points in width. Many graphics packages allow the user to specify a "hairline" for a very thin line. Hairlines often look acceptable when proofed on a typical laser printer. However, when produced on a high-resolution laser imagesetter, hairlines become nearly invisible and will be lost entirely in the final printing process.

Screens should be set to values between 15% and 85%. Screens which fall outside of this range are too light or too dark to print correctly. Variations of screens within a graphic should be no less than 10%.

Any graphics created in color will be rendered in grayscale for the printed version unless color printing is authorized by the Managing Editor and the Publisher. In general, color graphics will appear in color in the online version.

Inquiries. Any inquiries concerning a paper that has been accepted for publication should be sent to `memo-query@ams.org` or directly to the Electronic Prepress Department, American Mathematical Society, 201 Charles St., Providence, RI 02904-2294 USA.

Editors

SELECTED PUBLISHED TITLES IN THIS SERIES

1506 **Radu Saghin, Pancho Valenzuela-Henríquez, and Carlos H. Vásquez,** Regularity with Respect to the Parameter of Lyapunov Exponents for Diffeomorphisms with Dominated Splitting, 2024

1505 **Deniz Bilman and Peter D. Miller,** Extreme Superposition: High-Order Fundamental Rogue Waves in the Far-Field Regime, 2024

1504 **Daniele Alessandrini, Olivier Guichard, Eugen Rogozinnikov, and Anna Wienhard,** Noncommutative Coordinates for Symplectic Representations, 2024

1503 **J. Adams, N. Arancibia Robert, and P. Mezo,** Equivalent Definitions of Arthur Packets for Real Classical Groups, 2024

1502 **R. M. Guralnick and R. Lawther,** Generic Stabilizers in Actions of Simple Algebraic Groups, 2024

1501 **Erasmo Caponio, Miguel Angel Javaloyes, and Miguel Sánchez,** Wind Finslerian Structures: From Zermelo's Navigation to the Causality of Spacetimes, 2024

1500 **Kenji Fukaya, Yong-Geun Oh, Hiroshi Ohta, and Kaoru Ono,** Exponential Decay Estimates and Smoothness of the Moduli Space of Pseudoholomorphic Curves, 2024

1499 **Stephen Theriault,** Homotopy Fibrations with a Section after Looping, 2024

1498 **Fabio Pusateri and Avraham Soffer,** Bilinear Estimates in the Presence of a Large Potential and a Critical NLS in 3D, 2024

1497 **Itay Glazer and Yotam I. Hendel,** On Singularity Properties of Word Maps and Applications to Probabilistic Waring Type Problems, 2024

1496 **Hongyu He,** Unipotent Representations, Theta Correspondences, and Quantum Induction, 2024

1495 **Fabien Trihan and David Vauclair,** A Comparison Theorem for Semi-Abelian Schemes over a Smooth Curve, 2024

1494 **Francesco Fournier-Facio,** Normed Amenability and Bounded Cohomology over Non-Archimedean Fields, 2024

1493 **Javier Fernández de Bobadilla and Agustín Romano Velázquez,** Reflexive Modules on Normal Gorenstein Stein Surfaces, Their Deformations and Moduli, 2024

1492 **Dongxiao Yu,** Asymptotic Completeness for a Scalar Quasilinear Wave Equation Satisfying the Weak Null Condition, 2024

1491 **David Beltran, Joris Roos, and Andreas Seeger,** Multi-scale Sparse Domination, 2024

1490 **Jack Kelly,** Homotopy in Exact Categories, 2024

1489 **John Bergdall and David Hansen,** On p-Adic L-Functions for Hilbert Modular Forms, 2024

1488 **David J. Benson,** Modular Representation Theory and Commutative Banach Algebras, 2024

1487 **Paul Apisa and Alex Wright,** Reconstructing Orbit Closures from their Boundaries, 2024

1486 **M. Gekhtman, M. Shapiro, and A. Vainshtein,** A Plethora of Cluster Structures on GL_n, 2024

1485 **David Ayala, Aaron Mazel-Gee, and Nick Rozenblyum,** Stratified Noncommutative Geometry, 2024

1484 **Brandon Doherty, Krzysztof Kapulkin, Zachery Lindsey, and Christian Sattler,** Cubical Models of $(\infty, 1)$-Categories, 2024

1483 **Masao Oi,** Simple Supercuspidal L-Packets of Quasi-Split Classical Groups, 2024

1482 **David Burns and Daniel Macias Castillo,** On Refined Conjectures of Birch and Swinnerton-Dyer Type for Hasse–Weil–Artin L-Series, 2024

For a complete list of titles in this series, visit the
AMS Bookstore at **www.ams.org/bookstore/memoseries/**.